Vibhuti Narayan Rai

Light Emission from Single Self-decoupled Molecules in a Scanning Tunnelling Microscope

Experimental Condensed Matter Physics
Band 30

Eine Übersicht aller bisher in dieser Schriftenreihe erschienenen Bände finden Sie am Ende des Buchs.

Light Emission from Single Self-decoupled Molecules in a Scanning Tunnelling Microscope

by
Vibhuti Narayan Rai

KIT Scientific Publishing

Karlsruher Institut für Technologie
Physikalisches Institut

Light Emission fromSingle Self-decoupled Molecules
in a Scanning Tunnelling Microscope

Zur Erlangung des akademischen Grades eines Doktors der Naturwissenschaften von der KIT-Fakultät für Physik des Karlsruher Instituts für Technologie (KIT) genehmigte Dissertation

von Vibhuti Narayan Rai, M.Sc. Phys.

Tag der mündlichen Prüfung: 22. October 2021
Referent: Prof. Dr. Wulf Wulfhekel
Korreferent: Prof. Dr. David Hunger

Printed and/or published with the support of German Academic Exchange Service (DAAD).

Impressum

KIT Scientific Publishing

Karlsruher Institut für Technologie (KIT)
KIT Scientific Publishing
Straße am Forum 2
D-76131 Karlsruhe

KIT Scientific Publishing is a registered trademark
of Karlsruhe Institute of Technology.
Reprint using the book cover is not allowed.

www.ksp.kit.edu

Print on Demand 2022 – Gedruckt auf FSC-zertifiziertem Papier

ISSN 2191-9925
ISBN 978-3-7315-1201-1
DOI 10.5445/KSP/1000146586

Dedicated to My Parents "Arvind Kumar Rai" and late "Shail Rai"

List of abbreviations

CAD Computer-aided design

CCD Charge-coupled device

DFT Density functional theory

FWHM Full width at half maximum

HL Hot electroluminescence

HOMO Highest occupied molecular orbital

IET Inelastic energy transfer

ITE Inelastically tunnelling electrons

LDOS Local density of states

LSP Localized surface plasmon

LUMO Lowest unoccupied molecular orbital

NTO Natural transition orbital

OLED Organic light-emitting diode

PL Photoluminescence

SPP Surface plasmon polariton

STM Scanning tunnelling microscopy

STML Scanning tunnelling microscope induced luminescence

STS Scanning tunnelling spectroscopy

TERS Tip-enhanced Raman spectroscopy

TR-STML Time-resolved scanning tunnelling microscopy

UCEL Up-conversion electroluminescence

UHV Ultra-high vacuum

Contents

1 Introduction

Researchers have always been interested in studying objects which are not accessible directly via our senses, whether these objects are far beyond our galaxy or as small as a single atom. Over the last several centuries, the invention and evolution of telescopes and detector techniques have enabled the investigation of interstellar objects. However, in regard to the nanoscopic structures, most of the breakthrough discoveries were made after the introduction of quantum physics in the beginning of the 20th century. Since then, researchers have been able to investigate single atoms and molecules with the purpose to exploit various quantum mechanical phenomena for device applications. This has led to tremendously fast progress in the development of highly efficient and compact devices. Moreover, the current trend of miniaturization of devices suggests that the transistor sizes are already approaching the limit where quantum mechanical effects are no longer negligible. This means that the interest in alternative technologies based on controlling quantum mechanical phenomena is not only of scientific curiosity, but it is the demand for future electronic devices. One of the alternatives is to use molecules as the functional building blocks to combine their inherent properties with the conventional functionalities of existing technologies.

The possibility to convert electrical current into light, so called electroluminescence, is highly attractive, especially for the development of organic light emitting diodes (OLEDs). In the past few decades, the optical excitation of molecules at the nanoscale has revealed a variety of properties which are otherwise suppressed in ensembles [1]. In 1989, Moerner *et al.* [2] first reported the single molecule spectroscopy and subwavelength resolution imaging for which W. Moerner, E. Betzig, and S. Hell were awarded with Nobel Prizes in 2014. They studied single molecules within ensembles by using the wavelength selectivity. Even though these revolutionary work provides information about the various excited states of the molecules, determining the local electronic environment was not possible. Over the next few decades, among many innovative methods, the field of nanoplasmonics has emerged which managed to reduce the mismatch between the focused light field (several hundreds of nanometer in the visible range) and the extension of electron wavefunction in the matter with features measuring in sub-nanometerscale [3]. In nanophotonics, confinement of light below the Abbe diffraction limit [4] has allowed to study the light-matter interaction at the nanoscale.

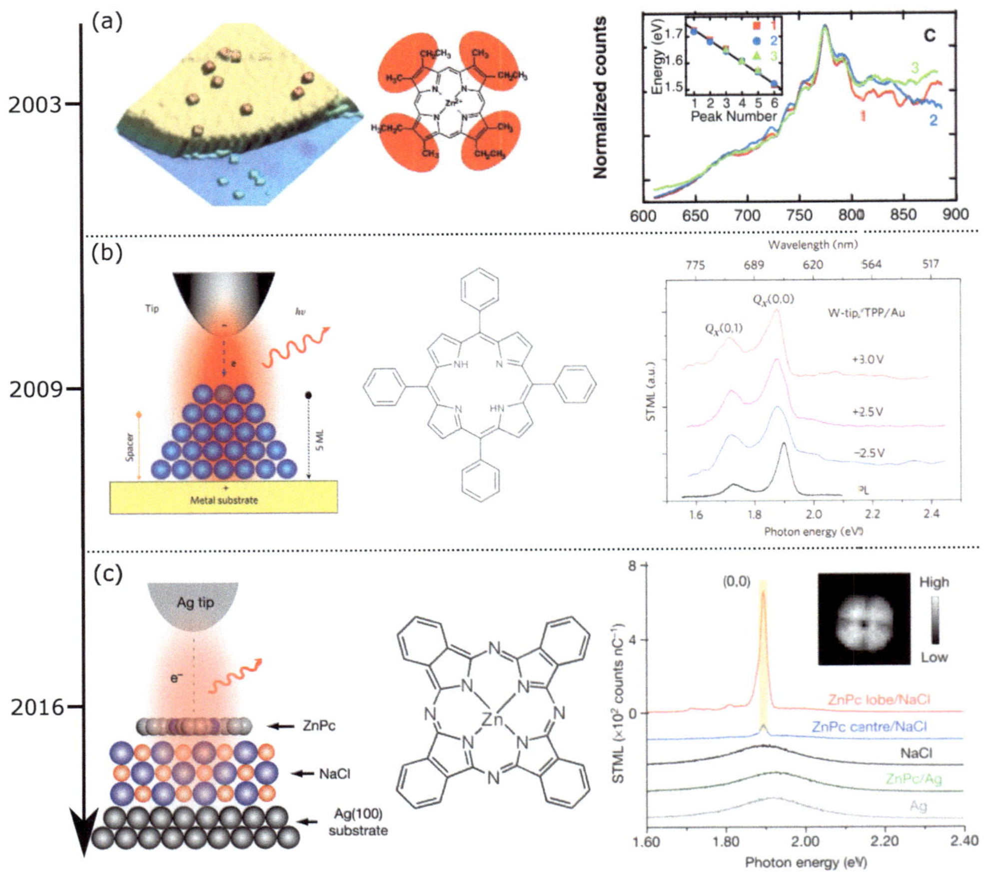

Figure 1.1: Different methods to decouple molecules from metal substrates. (a) Decoupling porphyrin molecules via thin layer of Al_2O_3 (adapted from [5]). **(b)** Multilayer of molecules for electronic decoupling [6]. **(c)** Decoupling ZnPc molecules by trilayer of NaCl [7]. Images in **b** and **c** are reprinted with permission from Ref. [6, 7].

There are several methods which can be used for the confinement of light below the diffraction limit. However, after the invention of scanning tunnelling microscopy (STM) [8, 9], it has been possible to analyze the spatial distribution of the electronic states responsible for light emission upon electrical charge injection. Nonetheless, studies of STM induced luminescence (STML) started in late 1980s with the pioneering work of Gimzewski *et al.* [10] and Berndt *et al.* [11] on single crystal metals and later extended to semiconductors and adsorbates [12, 13]. This really opened a door to study molecular luminescence with spatial resolution. However, this has been challenging because of two conflicting requirements: First, molecules need to be in contact with a conducting lead (substrate) to support quantum tunnelling in the STM junction and second, the molecule has to be electronically decoupled from the substrate to prevent fluorescence quenching. In the last two decades, several methods have been developed to decouple molecules. Basically these methods can be divided into two categories: First where the decoupling is achieved by inserting a thin insulating layer between the molecule and the substrate, whereas the second method involves having so-called self-decoupled molecules where a molecule has a functional unit which can bond to the substrate and make the chromophore part of the molecule electronically decoupled.

Figure 1.1 shows the advancements made in the last two decades to decouple molecules by using an insulating layer. The first successful implementation of this idea was reported by Qiu *et al.* [5] in 2003 (see Fig. 1.1a). They reported molecular fluorescence from individual porphyrin molecules adsorbed on an ultrathin alumina (Al_2O_3) film grown on a NiAl(110) surface. This boosted the field of electroluminescence from single organic/inorganic molecules. However, in this report, the observed molecular spectra did not match the fluorescence of the molecule, which means that the decoupling was not sufficient and the molecular states were hybridized with the substrate. In 2009, Dong *et al.* [6] reported decoupling by using a molecular multilayer but this approach lacked the single molecule resolution (see Fig. 1.1b). Soon after the report of Dong *et al.*, Zhang *et al.* (in 2016) reported well resolved fluorescence of single ZnPc molecules decoupled via a trilayer of NaCl on Ag(111) (see Fig. 1.1c) [7]. They were able to demonstrate that the highly localized excitations (plasmons) can be used to map the spatial distribution of the excitonic coupling in well-defined molecular adsorption geometry. Roughly at the same time, Imada *et al.* [18] reported the enrgy transfer between two Mg and H_2-phthalocyanine molecules by using the same method as Zhang *et al.*. Since then, this method has been employed to study various molecules and phenomena associated with them [3, 19–30].

In one of the earlier attempts to self-decoupled molecules, Schneider *et al.* [14] used a double-decker molecule comprising two separate π -electron systems (see Fig. 1.2a). The emission spectra showed some features related to the vibrational molecular states, however, the molecules were hybridized with the substrate and any sharp emission line of molecular origin was absent. Soon after that, Zhu *et al.* [15] reported the self-decoupling of porphyrin molecules by a tripodal anchor (see Fig. 1.2b). Some of the molecules showed sharp emission lines. However, the majority of molecules tend to adopt flat lying configurations. In 2016, Chong *et al.* [16] considered another approach where a single porphyrin molecule was suspended between the tip and the sample by molecular oligothiophene wires. This method resulted in vibrationally resolved sharp emission lines of molecular origin but lacked spatial resolution (see Fig. 1.2c). In 2019, Ijaz *et al.* [17] reported light emission from a fully symmetric tetrapodal molecule, among which three functional arms served as a tripod platform and the fourth arm was in the upward direction (see Fig. 1.2d). This was a very interesting approach since it insures

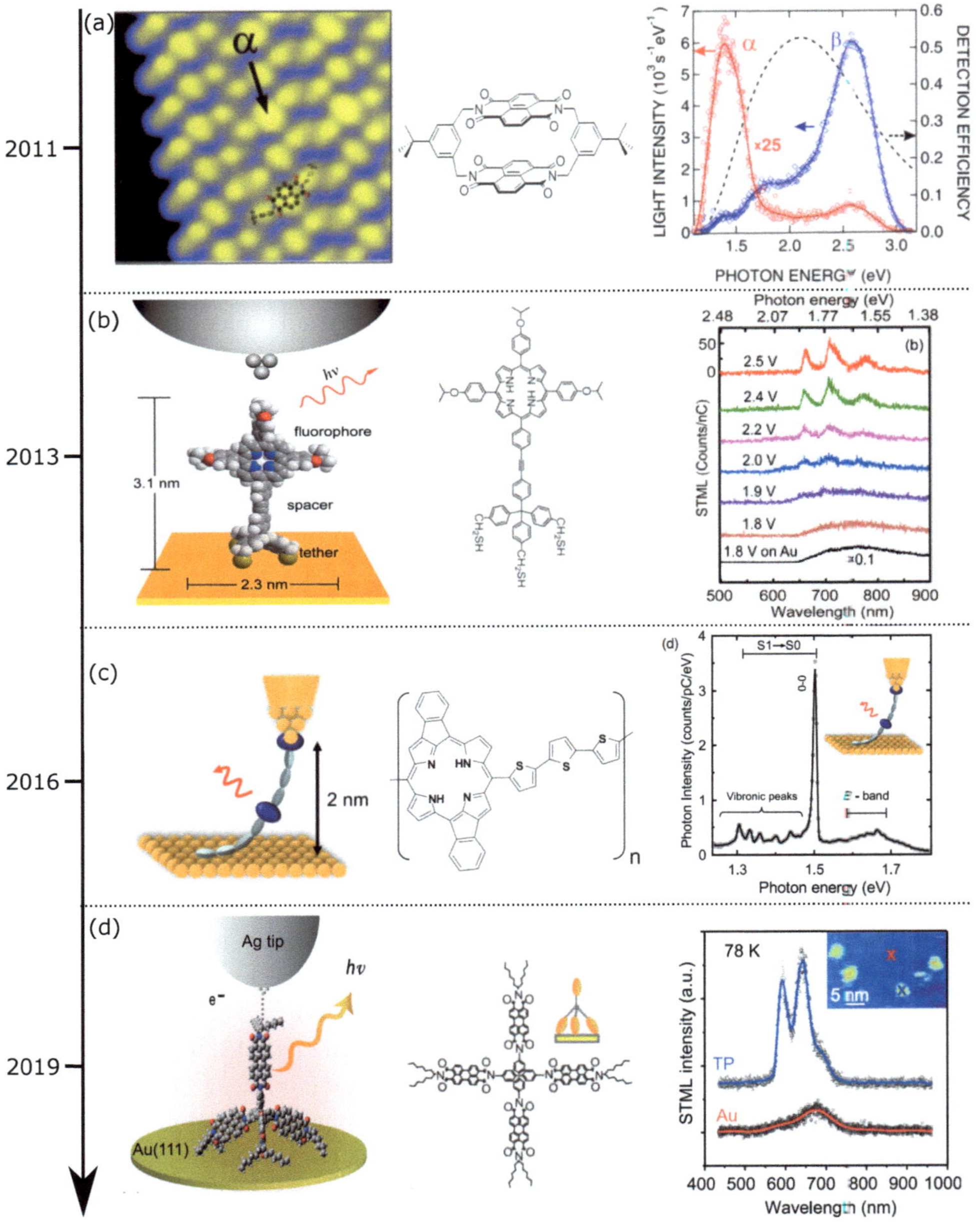

Figure 1.2: Reports on self-decoupling of molecules. (a) Double-decker molecular configuration [14]. **(b)** Tailor made self-decoupled molecule [15]. **(c)** Manipulating the chromophore by STM tip in a molecular wire [16]. **(d)** Another design for tailor made self-decoupled molecule [17]. Images are reprinted with permission from Ref. [14–17].

that at least one arm should be decoupled. However, the observed emissions were not fixed and varied in a large energy window. This clearly shows the complexity involved when dealing with single self-decoupled molecules.

The advancement of molecular optoelectronics depends on two key factors: **(1)** Understanding the mechanism behind efficient light emission from single molecules and **(2)** to control the luminescence of a single molecule directly bridging two metallic electrodes. The goal of the thesis is to understand and attempt to resolve the two forementioned issues. Consequently, this thesis is divided in six chapters (excluding the current chapter).

Chapter 2 will start with the discussion of the basic principles behind the working of STM and the theoretical basis. Next part of the chapter is dedicated to the discussion of the mechanism behind light emission from tunnelling junctions (nanocavities), followed by the discussion of possible phenomenological models to explain the light emission from single molecules in the STM junction. In the last part, our approach towards self-decoupled molecules will be presented. This includes the discussion of two new types of tripodal platforms, which are tailor-made for achieving the self-decoupling of light emitting chromophores.

Chapter 3 will discuss the experimental setup used for obtaining the results presented in this thesis.

Chapter 4 is the first among the three result chapters. In this chapter a comprehensive study of light emission from H_2Pc molecule will be discussed. We report the newly found emission line of H_2Pc and with the help of quantum chemical calculations, we successfully assigned it to its charged species. The results obtained on the charged and neutral molecules enable us to quantify their emission efficiencies. Results presented in this chapter were published in 2020 [29]

Chapter 5 will discuss the results obtained on three different molecules, Tpd-hNDI, Tpd-sNDI and Tpd-nNDI, designed for self-decoupling. The first part of the chapter will discuss the STM and STML measurements on these molecules in the pristine state. The second part of the chapter will discuss the influence of the STM tip on the adsorption geometry and emission spectra of Tpd-sNDI. Some of the results discussed in this chapter were published in 2021 [31].

Chapter 6 will discuss the results obtained on two chomophores -sNDI and -nNDI, anchored to a new type of tripodal platform (Tpdx). The design of this platform was realized based on the results obtained in the previous two chapters. STML results obtained on these molecules Tpdx-sNDI and Tpdx-nNDI suggest that the molecules stand in the upright orientation and have a larger component of the transition dipole moment perpendicular to the surface for efficient outcoupling of the light. In this chapter, a detailed discussion of the observed intriguing phenomena such as chemical modification of the molecules by the STM tip, vibrationally resolved optical emission, up-conversion electroluminescence, and hot electroluminescence will be discussed.

Chapter 7 will give a brief summary of the results and how they provide guideline to obtain an efficient single molecule light emitting circuit for future applications.

2 Theoretical and Experimental Background

In this chapter, a brief discussion of theories behind the experimental methods and observed phenomena will be discussed. The chapter is divided into three major parts: The first part discusses the basic principle behind STM and the approximations needed to understand the observations. In the second part, a brief discussion of the processes behind the light emission from the tunnelling junction will be discussed, followed by electroluminescence from organic molecules. Finally, the idea of molecular structures will be discussed, which are capable of emitting light in the tunnelling junction without suffering the fluorescence quenching.

2.1 Scanning Tunnelling Microscopy

In quantum mechanics, a particle has a nonzero probability of tunnelling through a potential barrier, which means the wavefunction of a particle does not vanish abruptly at the onset of a potential barrier. Instead decays exponentially with the distance [32]. This is the working principle of STM [9, 33]. In STM, two electrodes, an atomically sharp tip and the sample, are placed in close proximity (< 1 nm) and an electron (the particle) can tunnel from one electrode to another with finite probability by crossing the vacuum barrier. If a bias voltage is applied between the two electrodes, the tunnelling current gives information about the electronic structure of the sample.

Before going into the details of the theory of tunnelling, a brief description of the STM is needed. Figure 2.1 shows the schematic of STM with an atomically sharp tip placed on top of the sample. The tip is attached to piezoelectric transducers which can expand or contract with the application of voltage. In one mode of the STM operation, the distance between the tip and the sample is kept constant and tunnelling current is measured through vacuum. This is called the constant height mode of imaging. In another mode of operation, the tunnelling current is kept constant by changing the distance between the tip and the sample with the help of a feedback loop. The voltage required to adjust

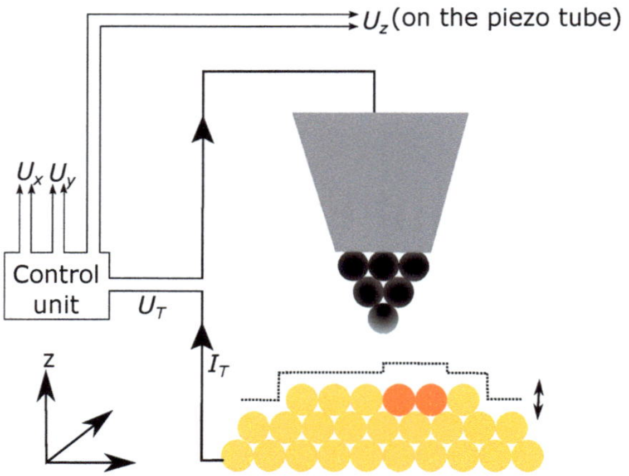

Figure 2.1: Schematic of tip-sample geometry connected with a control unit. The Voltages U_{xyz} are applied on the piezo tube for lateral and vertical motions. A bias voltage U_T is applied between the tip and the sample and the corresponding tunnel current (I_T) is measured. In the constant current mode, a feedback loop maintains a constant I_T by systematically changing the z-distance.

the piezoelectric crystal is then converted to the distance in order to get the topography. This is called constant current mode of imaging [34].

2.1.1 Quantum Tunnelling

To understand the working principle of STM, the discussion of the elementary theory of quantum tunnelling through a one-dimensional barrier is necessary.

An electron feeling a one dimensional potential $U(z)$ (see Fig. 2.2) is described by the wave function $\Psi(z,t)$, which satisfies the Schrödinger equation:

$$i\hbar\frac{\partial\Psi}{\partial t} = -\frac{\hbar^2}{2m}\frac{\partial^2\Psi}{\partial z^2} + U\Psi. \tag{2.1}$$

Solving the above equation by separation of variables for $\Psi(z,t) = \psi(z)\phi(t)$, the partial differential equation becomes:

$$-\frac{\hbar^2}{2m}\frac{d^2\psi(z)}{dz^2} + U\psi(z) = E\psi(z). \tag{2.2}$$

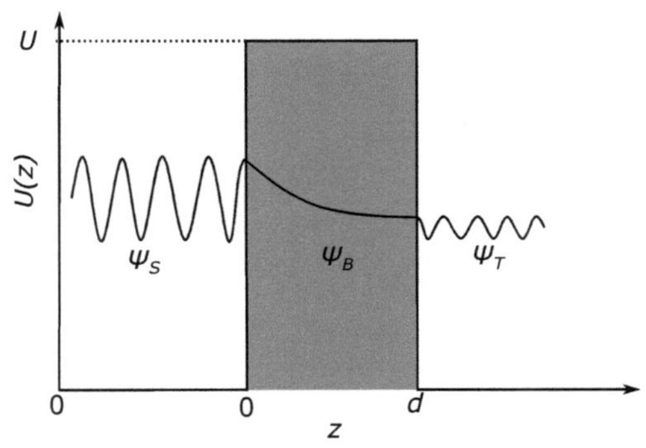

Figure 2.2: Schematic of the tunnel barrier with width d and potential U.

In this model, the distance between the STM tip and the sample is z, the vacuum gap is modeled by the potential barrier U (see Fig. 2.2) and $E = i\hbar\frac{1}{\phi}\frac{d\phi}{dt}$. When $E >\mid U \mid$, the solutions of eq. 2.2 are, $\psi(z) = \psi(0)e^{\pm ikz}$, where

$$k = \frac{\sqrt{2m(E-\mid U \mid)}}{\hbar}$$

is the wave vector. In the classically forbidden region, which means $E <\mid U \mid$, the solution is given by $\psi(z) = \psi(0)e^{-\kappa z}$ with

$$\kappa = \frac{\sqrt{2m(\mid U \mid -E)}}{\hbar},$$

where κ is the decay constant describing the electron penetrating through the barrier into the $+z$ direction. Electrons can also propagate in the opposite direction $(-z)$ after the reflection from the barrier. Therefore, the solution for a particle in every region, sample, barrier and tip are written as

$$\psi_S = e^{ikz} + Ae^{-ikz} \tag{2.3}$$

$$\psi_B = Be^{-\kappa z} + Ce^{\kappa z} \tag{2.4}$$

$$\psi_T = De^{ikz}, \tag{2.5}$$

where A and D are the reflection and transmission coefficients. Now imposing the boundary conditions of ψ and $d\psi/dt$ being continuous, the coefficients A, B, C and D can be obtained [32].

The incident (I_i) and transmitted (I_t) current densities can be obtained from the following equation:

$$I = \frac{\hbar}{2mi}\left\{\psi^*\frac{d\psi}{dz} - \psi\frac{d\psi^*}{dz}\right\} \tag{2.6}$$

Element	Al	Au	Cu	Ir	Ni	Pt	Si	W
Φ	4.1	5.4	4.6	5.6	5.2	5.7	4.8	4.8
κ	1.03	1.19	1.09	1.21	1.16	1.22	1.12	1.12

Table 2.1: Work functions and decay constant of different materials [35].

By using Eq. 2.3, 2.5, and 2.6, incident and transmitted current densities are:

$$I_i = \frac{\hbar k}{m}, \quad I_t = \frac{\hbar k}{m} \mid D \mid^2 . \tag{2.7}$$

The transmission probability through the barrier is given by the ratio of transmitted and incident current density:

$$T = \frac{I_t}{I_i} = \mid D \mid^2 = \frac{1}{1 + \frac{(k^2+\kappa^2)^2}{4k^2\kappa^2} \cdot \sinh^2(\kappa d)}, \tag{2.8}$$

where d is the gap between the tip and the sample. For a strongly attenuated barrier, the decay constant κ is large

$$T = \frac{16\kappa^2 k^2}{(k^2 + \kappa^2)^2} \cdot e^{-2\kappa d}. \tag{2.9}$$

Therefore, the tunnelling current exponentially decays as

$$I_T \propto T \approx e^{-2\kappa d}. \tag{2.10}$$

This basic model already gives the dependency of the tunnelling current, which decreases exponentially with the tunneling barrier thickness (d). This has been observed in STM experiments as well as in other experimental techniques [36–38]. This responsiveness of the tunnelling current to d explains the high sensitivity of the tip to the height changes in the topography. The magnitude of the decay constant depends on the barrier height. In STM, the barrier height corresponds to the metal work function (Φ), *i.e* the energy necessary to remove an electron from the bulk to the vacuum (in metals, the tunnelling electrons lie close to the Fermi energy). For a typical value of the work function ($\sim 5eV$), the decay constant (κ) is of the order of 1 Å. This means the tunnelling current decreases by one order as the tip-sample distance increases by 1 Å. Table 2.1 shows the typical values of the work functions of some materials which are commonly used in the STM experiments [35].

2.1.2 Tunnelling Current Approximations

In the above treatment, an essential fact was missing that the electrons are fermions and while tunnelling from one electrode to another, it is necessary to take into account the initial and the final quantum states of the electron. This occupation probability depends on the temperature and it is given by Fermi-Dirac distribution. To address these issues, Bardeen proposed a model, also referred as the transfer Hamiltonian method. This model was later extended by Tersoff and Hamann and again by Chen for STM [39–41]. In Bardeen's model, the following assumptions were considered: First, electron-electron interactions were neglected, which is a reasonable assumption in the low tunnelling regime. Furthermore, tip-sample coupled modes were not taken into account. Later in this chapter, the formation of coupled electronic modes will be discussed, which is necessary for the light emission from tunnelling junctions. In the following subsections, Bardeen as well as Tersoff and Hamann models will be discussed in brief [42].

Bardeen Model

If the two electrodes, tip (T) and the sample (S) are far apart, then the Schrödinger equation can be written as:

$$i\hbar\frac{\partial \Psi^i}{\partial t} = -\frac{\hbar^2}{2m}\frac{\partial^2 \Psi^i}{\partial z^2} + U_i\Psi^i, \tag{2.11}$$

where U_i ($i = S$ or T) is the potential and Ψ^i is the wavefunction, which depends on both, time and spatial coordinates. The stationary states can be written as $\Psi^i = \psi^i_\mu\, e^{\frac{-iE^i_\mu t}{\hbar}}$ and similar to the eq. 2.2 is:

$$-\frac{\hbar^2}{2m}\frac{d^2\psi^i_\mu}{dz^2} + U_i\psi^i_\mu = E^i_\mu\psi^i_\mu. \tag{2.12}$$

As the tip and the sample are brought closer, the Schrödinger equation for the coupled system contains the total potential:

$$i\hbar\frac{\partial \psi}{\partial t} = \left(-\frac{\hbar^2}{2m}\frac{\partial^2}{\partial z^2} + U_S + U_T\right)\psi. \tag{2.13}$$

Initially the tip is far, which means the electron is in a state ψ^S_μ of the sample. Now the tip is approached and the potential is turned on adiabatically and the time-dependent potential can be written as:

$$U_T(t) = e^{\beta t/\hbar}U_T, \quad \beta > 0, \tag{2.14}$$

where U_T is constant. As the time progresses, ψ_μ^S is not described by eq. 2.12 instead has a probability to populate the states of the tip (ψ_ν^T). This probability is included in the tunnelling current. The wavefunctions of the tip or the sample are not the eigenfunction of the Hamiltonian of the combined system. Now the wavefunction of the combined system ψ can be written as linear combination of the sample and the tip eigenfunctions:

$$\psi = a_\mu(t)\psi_\mu^S e^{-iE_\mu^S t/\hbar} + \sum_{\nu=1}^{\infty} c_\nu(t)\psi_\nu^T e^{-iE_\nu^T t/\hbar}. \tag{2.15}$$

The coefficients a_μ (t) and c_ν (t) are 1 and 0 initially. In Bardeen's theory, the two sets of wavefunctions (ψ_μ^S and ψ_ν^T) are orthogonal, which means

$$\langle \psi_\nu^T \mid \psi_\mu^S \rangle \approx 0. \tag{2.16}$$

After solving eq. 2.13 for the combined wavefunction ψ (from eq. 2.15),

$$i\hbar \frac{dc_\nu(t)}{dt} = \langle \psi_\nu^T \mid U_T \mid \psi_\mu^S \rangle e^{-i(E_\mu^S - E_\nu^T + i\beta)t/\hbar} + \ldots\ldots \tag{2.17}$$

In the above solution, higher order terms are neglected and due to the adiabatic approximation, $a_\mu(t) = 1$. Here, the integral $\langle \psi_\nu^T \mid U_T \mid \psi_\mu^S \rangle$ is the matrix element $M_{\mu\nu}$. Integrating eq. 2.17 gives:

$$c_\nu(t) = \frac{1}{(E_\mu^S - E_\nu^T + i\beta)} M_{\mu\nu} e^{-i(E_\mu^S - E_\nu^T + i\beta)t/\hbar}. \tag{2.18}$$

The probability of populating the state ψ_ν^T at time t from the initial state ψ_μ^S is given by

$$\mid c_\nu(t) \mid^2 = \frac{e^{2\beta t/\hbar}}{(E_\mu^S - E_\nu^T)^2 + \beta^2} \mid M_{\mu\nu} \mid^2 . \tag{2.19}$$

The time derivative of the above equation gives tunnelling probability per unit time and by using the definition of δ function (in limit $\beta \to 0$):

$$P_{\mu\nu}(t) = \frac{2\pi}{\hbar} \delta(E_\mu^S - E_\nu^T) \mid M_{\mu\nu} \mid^2 . \tag{2.20}$$

This is Fermi's Golden Rule. The δ function in eq. 2.20 takes care of the elastic tunnelling. The tunnelling current is determined by $eP_{\mu\nu}$. Up to this point, the argument of finite temperature has not been considered. So, at a finite temperature, occupied and unoccupied states are described by Fermi-Dirac distribution as $f(E - E_F) = (1 +$

$e^{[(E-E_F)/k_BT]})^{-1}$ and $1-f(E-E_F)$, respectively. Therefore, if the applied bias voltage is V then the tunnelling current is:

$$\begin{aligned} I &= \frac{4\pi e}{\hbar}\int d\epsilon[f(E_F^T - eV + \epsilon) - f(E_F^S + \epsilon)] \\ &\times \rho^T(E_F^T - eV + \epsilon)\rho^S(E_F^S + \epsilon) \mid M(E_F^S + \epsilon, E_F^T - eV + \epsilon) \mid^2, \end{aligned} \tag{2.21}$$

where ρ^T and ρ^S are the local density of states (LDOS) of the tip and the sample. This directly shows that the tunnelling current depends on the LDOS of the tip and the sample. Bardeen's tunnelling matrix element can be written as:

$$M_{\mu\nu} = \frac{\hbar^2}{2m}\int_{\Sigma}(\psi_\mu^S\vec{\nabla}\psi_\nu^{T*} - \psi_\nu^{T*}\vec{\nabla}\psi_\mu^S)\cdot d\vec{S}. \tag{2.22}$$

This surface integral is to be evaluated over the surface within the barrier region. Bardeen's model was originally meant to describe the metal-insulator-metal junction transport. Later, Tersoff and Hamann applied this method to STM [40].

Tersoff-Hamann Model

The Bardeen model showed that the tunnelling current is a convolution of the electronic state of the tip and the sample, however, the calculation is very complex to actually compute the tip electronic structure. With the purpose to simplify Bardeen's model, Tersoff and Hamann proposed to consider the tip as a geometrical point and the state is spherically symmetric (s-wave).

If the tip spherical potential well is centered at $\vec{R_T}$, then the tunnelling current depends on the DOS of the sample at the tip position is given as:

$$I(\vec{R_T}, d, V) \propto \int_0^{eV} \rho^T(E_F^T - eV + \epsilon)\rho^S(\vec{R_T}, E_F^S + \epsilon)T(\epsilon, d, V)d\epsilon, \tag{2.23}$$

where T is the transmission coefficient which depends on the shape of the barrier. At low temperatures, the above equation:

$$I(\vec{R_T}, d, V) \propto \int_0^{eV} \rho^S(\vec{R_T}, E_F^S + \epsilon)T(\epsilon, d, V)d\epsilon. \tag{2.24}$$

This model has been proven to be very successful interpreting the characteristic features such as superstructure of surface reconstruction, defects, scattered waves, adsorbates, etc. However, it has its limitations since the electronic states of the tip are also involved in the observations.

2.1.3 Scanning Tunnelling Spectroscopy

The derivative of the tunnelling current with respect to the bias voltage gives the differential conductance and it is proportional to the local density of states of the sample:

$$\frac{dI}{dV} \propto \rho^S(\vec{R_T}, E_F^S + \epsilon). \tag{2.25}$$

In the experiment, I-V curves are recorded and the derivative dI/dV gives the information about the LDOS. Another, more subtle way to directly measure the dI/dV spectrum is by using a lock-in amplifier. Therefore, the voltage (V_0) is modulated by adding a small oscillating signal ($V^{'}$) of frequency ν. The resulting tunnelling current is then demodulated and it is proportional to dI/dV (see equation 2.26). Moreover, with lock-in amplifier, signal-to-noise ratio is much better.

$$I(V_0 + V' \sin(2\pi\nu t)) = I(V_0) + \left.\frac{dI}{dV}\right|_{V_0} V' \sin(2\pi\nu t) + \text{higher order terms} \tag{2.26}$$

2.1.4 Inelastic Tunnelling

Up to this point, every formulation considered that the energy of the electrons is the same after each tunnelling event, which is called elastic tunnelling process. However, tunnelling electrons can lose some of their energy in the junction and can trigger different excitations [43–45]. These excitations are observed as vibrations (phonons) [46], photons [5, 47], or spin-excitations [48], as shown by a schematic in Fig. 2.3a [45]. The tunnelling electrons in the junction can couple to the excitation mode and additional tunnelling channels can be created. This increases the tunnelling probability, which reflects in the measured tunnelling current as an increase. Figure 2.3b (top panel) shows the I-V curve where at the energy required for an excitation ($\hbar\omega$), tunnelling current suddenly increases with different slope. The derivative of such a curve gives an step function in the spectrum (see Fig. 2.3b, bottom panel).

Similar to the elastic contribution (eq. 2.20), the inelastic probability rate can be written as,

$$\delta P_{\mu\nu}(t) = \frac{\pi}{2\hbar}\delta(E_\mu^S - E_\nu^T \pm \hbar\omega) \mid \delta M_{\mu\nu} \mid^2 . \tag{2.27}$$

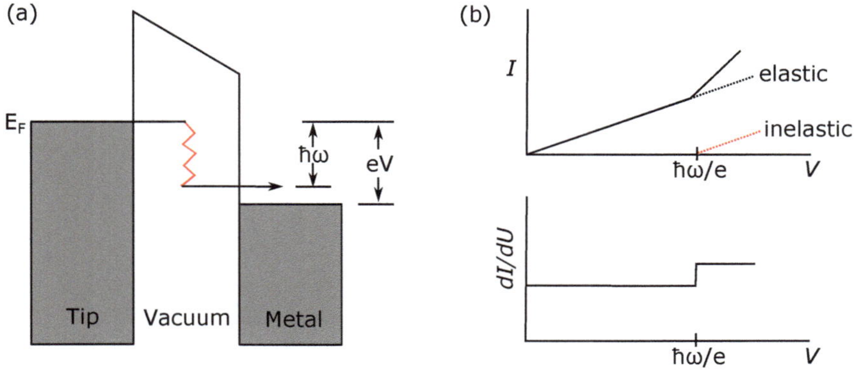

Figure 2.3: **(a)** Schematic representation of inelastic tunnelling (Ho *et al.* [45]), where $\hbar\omega$ is the energy of the excitation triggered by inelastically tunnelling electrons. **(b)** Tunnelling current (top) and conductance (bottom) as a function of applied bias.

2.2 Scanning Tunnelling Microscope Induced Luminescence

In the previous section, it was discussed that inelastic tunnelling electrons can excite different process. In the STM, the tip and the sample form a junction/nanocavity. In this nanocavity, a collective charge oscillation exists and can be excited by the inelastic tunnelling electrons. In turns, light is emitted from the junction into the far-field. The emitted light is a measure of the inelastic tunnelling electrons in the junction. This effect was first observed in a metal-insulator-metal junction by Lambe *et al.* in 1976 [47]. Later, in 1988, Gimzewski *et al.* observed this phenomenon in an STM junction [10] and opened a new way for various experiments by using scanning tunnelling microscopy induced luminescence (STML).

An STM junction has a confined electric field between two electrodes, which means the characteristics of the observed emission spectra depends on the involved material of the tip/sample and the confinement of the field, that is the geometry of the junction/nanocavity. The effect of the tip/sample material was studied by Berndt *et al.* for different metals [11]. The observed spectra showed broadband peaks centered at different energies for different materials. In the scope of this thesis, a similar measurement was carried out as shown in Fig. 2.4. Later, Meguro *et al.* reported a drastic difference in the observed STML spectra for different tip shapes [49].

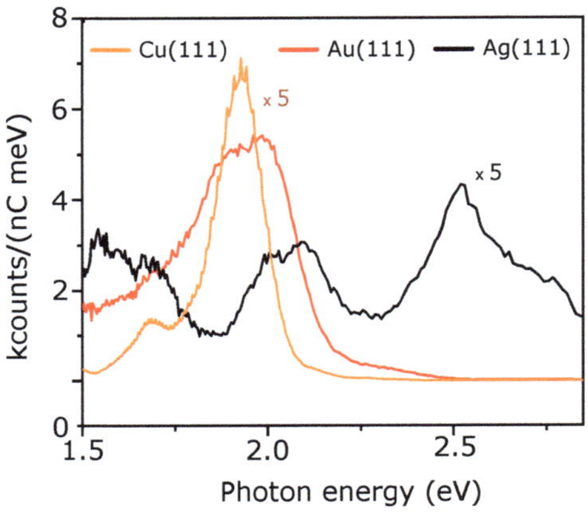

Figure 2.4: Optical emission spectra recorded on different substrates with a Ag tip (U = 3.0 V, I(Ag) = 500 pA, I(Au) = 300 pA, I(Cu) = 300 pA, t = 1 s).

2.2.1 Surface Plasmon Polaritons

Before going into the detailed description of the light emission processes from an STM junction, it is important to first discuss the electromagnetic (EM) response of the metal surfaces. Conduction electrons on the metal surface move freely with a background of positively charged atom cores. Once subjected to electric field ($\mathbf{E}(t) = \mathbf{E}_0 \exp(-i\omega t)$), these electrons start to oscillate. From the solution of the equation of motion, the oscillation frequency, also called plasma frequency, is

$$\omega_p = \sqrt{\frac{ne^2}{\epsilon_0 m}}, \tag{2.28}$$

where n is the charge carrier density, m is the electron mass, and ϵ_0 is the dielectric constant of the vacuum. The waves are quantatized and have an associated quasiparticle called plasmon (or bulk plasmon) with energy $\hbar\omega$ [50]. The penetration depth of the incident field is the skin depth of the metal $\delta = \sqrt{2/(\omega\mu_0\mu_m\sigma)}$, with ω is the incident wave, μ_0 is the magnetic permeability of the free space, μ_m is the relative metal permeability and σ is the metal conductivity [50]. There is no propagation of EM waves below the plasma frequency ($\omega < \omega_p$) whereas if $\omega > \omega_p$, waves propagate with a group velocity of $v_g = d\omega/dk$ [50].

These collective oscillations generated by the field are localized at the surface. At the interface between a metal (ϵ_m) and a dielectric (ϵ_d), the solutions of the Maxwell's equation in the two respective media give an exponentially decaying evanescent wave in the two half spaces that propagates along the interface. This coupled mode of charge

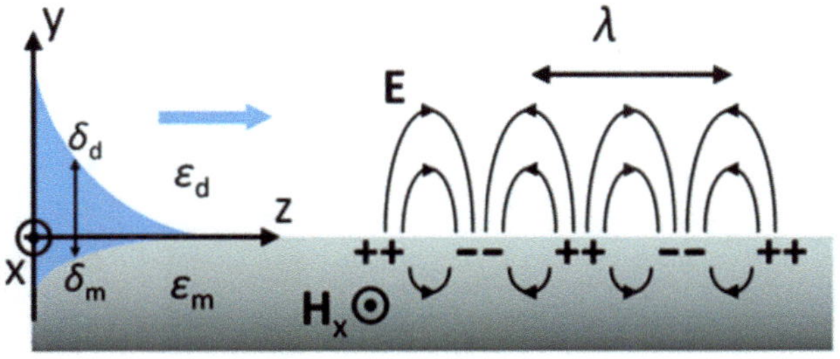

Figure 2.5: Schematic showing field components of a SPP at metal–dielectric interface, with components normal to the interface (y-axis) and along the propagation direction (z-axis). λ is the wavelength of propagating SPP (adapted from [53]).

carrier oscillation and the exponentially decaying field in the metal and dielectric is called surface plasmon polariton (SPP) [50–52]. A sketch is shown in Fig. 2.5 [53].

The SPP dispersion relation is:

$$k(\omega) = k_0(\omega)\sqrt{\frac{\epsilon_m(\omega)\epsilon_d(\omega)}{\epsilon_m(\omega)+\epsilon_d(\omega)}}, \tag{2.29}$$

and the penetration depth of the SPP in the two medium is

$$\delta_{d/m} = \frac{2\pi}{\lambda}\sqrt{\frac{\epsilon_m(\omega)+\epsilon_d(\omega)}{\epsilon_{d/m}^2}} \tag{2.30}$$

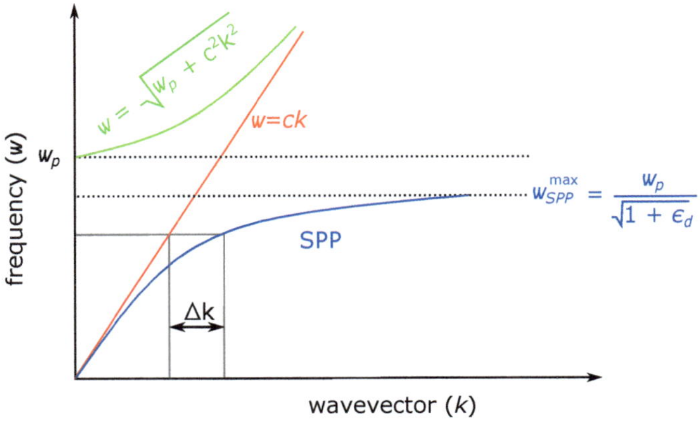

Figure 2.6: Dispersion relation of SPP (blue line), free propagating EM field (red line) and transverse optical modes of EM field in plasma (green line). Δk is the mismatch in the wavevector.

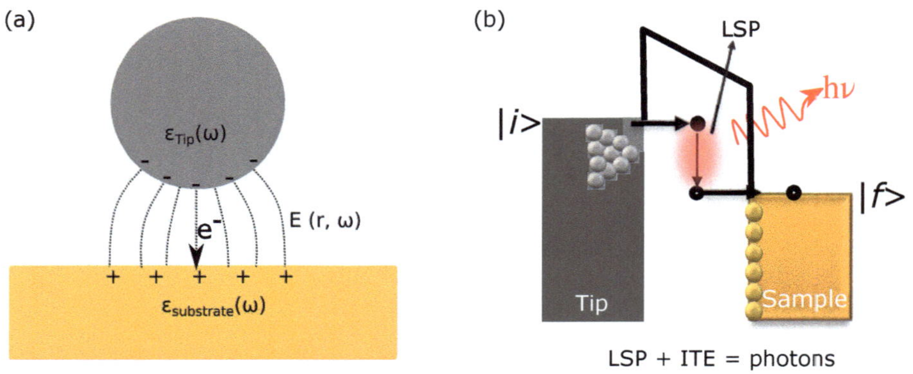

Figure 2.7: **(a)** Confined EM field in an STM cavity. **(b)** Schematic showing that the inelastically tunnelling electrons (ITE) exciting the LSPs.

It can be realized that SPP dispersion relation completely resides below the light cones ($\omega = ck$), while a crossing only occurs in the limit of $\omega = 0, \beta = 0$. Thus, SPPs cannot be excited by the direct illumination of light as the energy and momentum conservation cannot be fulfilled (see Fig. 2.6). Therefore, to excite the SPPs in order to decay radiatively, a momentum matching is necessary. This can be achieved by several methods for example, by prism coupling, grating coupling, or, as in the scope of the thesis, by the STM tip [3, 52–54].

2.2.2 Excitation of Gap Plasmons

Once two electrodes are approached close to each other, such as the tip and the sample in case of the STM (or two metal nano particles), the exponentially decaying evanescent waves overlap and form a coupled mode called localized surface plasmon (LSP) or gap plasmon [51] (Fig. 2.7a). It is clear that the SPPs and the LSPs are different in nature. The SPPs are waves propagating along the interface and cannot decay radiatively, whereas LSPs are localized in the STM junction and do not propagate. Moreover, LSPs can decay radiatively into the far-field. The LSP modes are dependent on the shape and the size of the tip-sample junction/nanocavity. In the simple geometry of the STM, where a sharp tip of radius R is placed in front of a flat sample at a distance d, Rendell *et al.* [55] calculated the field confinement in the junction and found that it scales as $\sqrt{2Rd}$. The distance between the tip and the sample is about 1nm which is far less than the wavelength of the emitted light, making STML a great tool to obtain lateral resolution with luminescence.

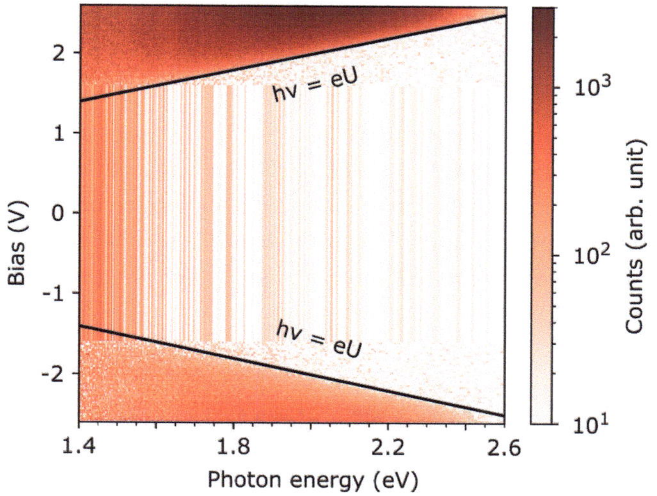

Figure 2.8: 2D color map of optical emission spectra recorded on Au(111) with a silver tip in the voltage range of $U = 2.60\,\mathrm{V}$ to $-2.60\,\mathrm{V}$ ($I = 30\,\mathrm{pA}$, $t = 3\,\mathrm{s}$). The black solid lines show the condition $h\nu = eU$.

The oscillating mode has resonances ω_l and in the limit $d << R$, it can be written as

$$\omega_l \approx \omega_p \sqrt{\left(l + \frac{1}{2}\right)\sqrt{\frac{d}{2R}}}. \tag{2.31}$$

This suggests that the tip shape is a dominant factor in deciding the shape of the recorded spectra. Moreover, as the tip approaches the sample, the resonances are red shifted [56, 57]. However, this classical picture does not fully describe the observations as the shift in the peak does not diverge at zero distance ($d = 0$) [58, 59].

Before moving further into the discussion of the characteristics of the light emission spectra obtained from the STM junction, it is important to obtain an intuitive picture of the processes leading to light emission. In the junction, LSPs can be excited by inelastically tunnelling electrons (ITE) and in turn can decay by emitting photons into the far-field, see the schematic in Fig. 2.7b. The energy of the emitted photon ($h\nu$) is limited by the difference in the energies of the initial ($| i\rangle$) and final ($| f\rangle$) states of the electron (see Fig. 2.7b). The LSPs eigenvalues form a continuum, which gives broad band spectra. Figure 2.8 shows a map of 82 spectra recorded in the bias range of 2.60 V to -2.60 V. The black lines are the values corresponding to $h\nu = eU$, which suggests the fact that the energy of the emitted photons is limited by the applied bias voltage. Light emission from such a metallic junction can be modeled as a dipole located in a STM junction with axis perpendicular to the sample. This was shown in the calculations by Rendell *et al.* [55]. In this condition, Fermi's golden rule in the dipole approximation can give the rate of emission [32, 60]:

$$\Gamma_{rad} = \frac{2\pi}{\hbar^2}|\langle i|\mu|f\rangle|^2 \rho(\omega_{if}) E_z(M)^2, \tag{2.32}$$

where $\langle i|\mu|f\rangle$ is the transition dipole moment, $\rho(\omega_{if})$ is the spectral density of EM modes and $E_z(M)$ is the local zero-point rms electric field at the location M of the quantum system. This means that the emission rate depends on two factors: a "source" in terms of the dipole moment operator and a "receptor" in terms of local and spectral density of the excited EM eigenmodes (the LSP modes), where $\rho(\omega_{if})$ represents the frequency dependent Purcell effect [61].

2.3 Light Emission from Single Molecules

An optical cavity enhances the light-matter interaction by trapping the light into a small volume, which increases the field. Another factor that leads to enhanced light-matter interaction is the quality factor, which is the time light spends in the cavity before escaping. Typically this is achieved in different types of optical cavities such as optical mirrors, dielectric resonators, and photonic crystals. However, all of these methods are limited by Abbe diffraction limit [4] on decreasing the effective volume. This issue can be overcome using a plasmonic nanocavity. In an STM junction, a very well-defined plasmonic nanocavity, the tip-sample junction is already present. This is very interesting because the STM not only provides a well-defined junction, but the optical excitations can be triggered in a very controlled manner by the tunneling current and applied bias voltage.

Over the last three decades, the experimental STML studies on single molecules have shown continuous and fast progress. These experiments have demonstrated many intriguing phenomena at the single molecule level [5–7, 19, 21, 23, 24, 29, 30, 62]. These studies reveal that mainly two excitation mechanisms are involved in the light emission processes. Mechanism I involves molecular excitation by electron-hole pair recombination [5, 18, 63, 64], whereas mechanism II suggests molecular excitation via plasmons [6, 16, 65]. In the following subsection, both mechanisms are discussed.

2.3.1 Excitation Mechanisms

Figure 2.9a shows the energy diagram of mechanism I for the light emission from a molecule in a tunnel barrier. At sufficiently high bias voltage, the Fermi energy of the tip is in resonance with occupied frontier molecular orbital (or HOMO). An electron is extracted from the molecule leaving a hole behind (process-1). If the system remains

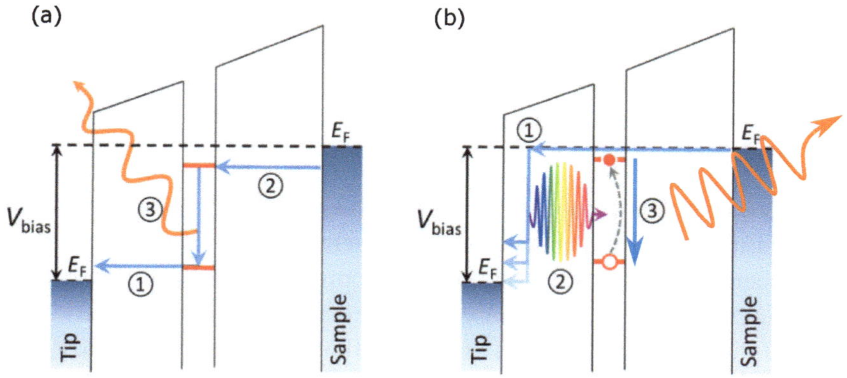

Figure 2.9: Energy diagram showing the excitation of a molecule in an STM junction via **(a)** injection of electron/hole from sample/tip and **(b)** via plasmons (IET). Image is reprinted from Ref. [3].

in this charged state for a sufficiently long time, the unoccupied frontier molecular orbital is then filled with another electron from the sample (process-2), this way an electron-hole pair is formed which can decay by recombination and emit a photon into the far-field (process-3). Optical emissions based on this mechanism require more energy than the energy of the emitted photons ($eU > h\nu$) because an extra energy is required to charge the molecule at the first place.

Figure 2.9b shows the energy diagram of mechanism II where the molecule is excited via plasmons. In this mechanism, an inelastically tunnelling electron (process-1) excites the LSPs (process-2). These LSPs act as a near-field light source, which creates excitons and in turn light is emitted into the far-field. This mechanism is observed for most of the STML studies on single molecules and is referred to inelastic energy transfer (IET) mechanism. It is important to mention here that the plasmon modes are involved in the efficient coupling of the molecular emission to the far-field, which means even if the observations support the electron-hole pair recombination mechanism, the effect of plasmons cannot be excluded.

2.3.2 Electronic Decoupling of Molecules

The two mechanisms discussed in the last section depend on the fact that molecule lives long enough in the excited metastable state such that it decays radiatively into the far-field. Therefore, light emission requires that the molecule should be electronically decoupled from the metal substrate to avoid fluorescence quenching. If the molecule is directly adsorbed on the metal substrate, the highest impedance is offered by the vacuum gap, whereas the molecular states are hybridized with the sample states. This hybridization results in a high decay rate (k), which is inversely proportional to the lifetime ($\propto 1/\tau$) of the excited state. Therefore, molecules decay non-radiatively and hence quenching of the fluorescence is observed [13, 66–71]. To overcome this issue, several methods have been employed over the years. These methods and their outcome were discussed in the chapter 1.

This thesis focuses on the decoupling of the molecules by virtue of tailor-made molecules for light emission. These molecules have a light-emitting chromophore anchored to a suitable spacer group (a tripodal platform) that preferentially binds to the metal and decouples the chromophore. So far, very few reports are available on such self-decoupled systems and the reason for this lies in the fact that these molecules have a complex nature and the adsorption geometry is less deterministic. The success of this approach lies on the design of the spacer group and the preparation of the sample. A major part of this thesis will discuss the optimal design of such tripodal platforms to achieve well reproducible light emission from single molecules.

2.4 Self-decoupled Molecular Structures for Light Emission

Figure 2.10a-e show the chemical structure of the molecules used for the STML experiments. Results on these molecules will be discussed in chapter 5 and 6. The molecules consist of various derivatives of naphthalene-1,4,5,8-tetracarboxdiimides (NDIs) as chromophores and they are anchored to two different kinds of tripodal platforms via an alkyne linker.

In the first kind of tripodal platform (Fig. 2.10a-c), chromophores are anchored to triphenylmethane platform with three acetyl-protected thiol anchors in the *meta-* position. The acetyl protection is cleaved off once the substrate is annealed at high temperatures. Triple bond sp^3 -hybridised carbon atoms provide rigidity for the STM measurements. The three aromatic thiol anchors allow direct connection to the Au substrate (Au-S bond after the deprotection of acetyl groups). The linker is expected to provide enough electronic decoupling to the chromophore from the substrate. It was observed that even though the chemistry of the molecule suggests that the molecules should bond to the Au(111) surface and stand upright, however, the STM and STML experiments show that the molecules lie flat on the surface and the fluorescence is quenched (see chapter 5).

Figure 2.10d and e show an improved design of the tripodal platform, which is expected to overcome the van der Waal force. Results in chapter 6 show that the tripodal platform is able to electronically decouple the chromophore and light emission from single

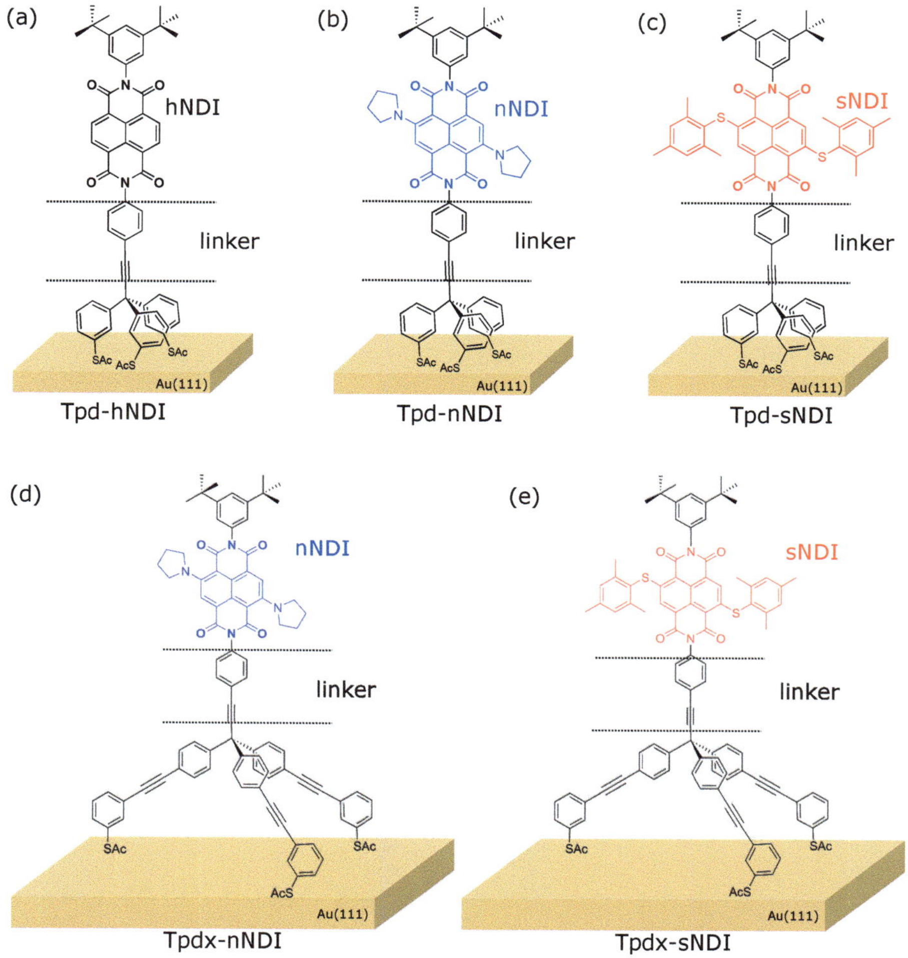

Figure 2.10: Chromophores anchored to tripodal platforms. (a)-(c) Three chromophores, -hNDI, -nNDI and -sNDI anchored to triphenylmethane platform. **(d)** and **(e)** -nNDI and -sNDI chromophores anchored to the extended platform.

molecules in an STM junction. A detailed discussion of the synthesis of complexes, including the platforms can be found in the PhD thesis of Nico Balzer [72].

3 Experimental Setup

In this chapter, an overview of the experimental setup is presented. We start with the discussion of UHV setup necessary for the preparation of impurity free atomically clean samples, followed by the discussion of the STM setup with optical access. Finally, different methods used for sample preparation will be discussed, which include cleaning of the sample and deposition of the molecules. The STM setup with optical access is homebuilt and designed in the group of Prof. Wulf Wulfhekel at the *Institute of Nanotechnology*, KIT [73, 74].

3.1 UHV Setup

For the experiments with single molecules, it is necessary to avoid impurities coming from the gas phase, which means performing experiments in UHV conditions. Thereby, typical base pressure for the sample preparation and STM measurements is maintained at 10^{-10} mbar, which means contamination of 1 molecule (Hydrogen) /nm^2 per hour. The basic layout of the experimental setup is shown in Fig. 3.1. The setup consists of three vacuum chambers (load-lock, STM, and preparation) separated by gate valves. Each of the chambers are first pumped with a rotary vane pump down to the pressure of $\sim 10^{-2}$ mbar. The rotary pump line also has a barrel to provide a buffer volume. Three turbo molecular pumps (TMPs) are connected to each of the chambers, which then create a UHV of $\sim 10^{-10}$ mbar. During the STM measurements, all pumps are turned off and the pressure is maintained by two ion getter pumps connected to the STM and preparation chambers. The ion getter pump in the preparation chamber also has titanium sublimation pump which helps to remove the residual gas and improve the UHV conditions.

The load-lock chamber is mainly used for transferring samples and STM tips from ambient conditions into the UHV. Two metal evaporators (Cr and Ag) are mounted on this chamber which were used to metalize the mirror tips (detailed discussion in section 3.2.2). An extended volume is connected to the chamber for spray deposition of organic molecules from solution. The pressure in the load-lock chamber is monitored by

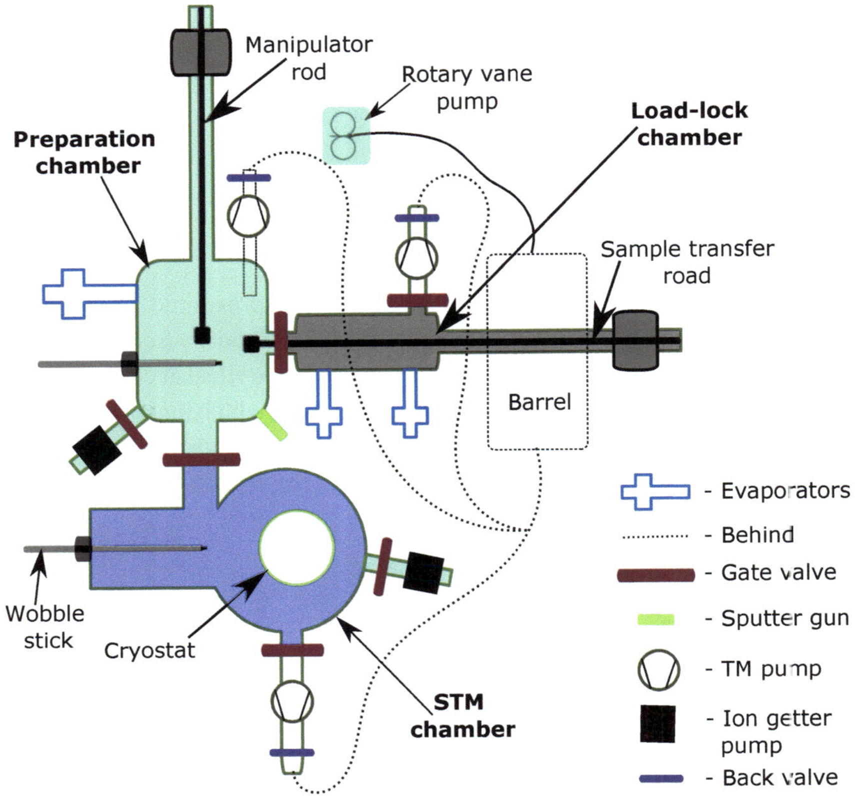

Figure 3.1: Schematic of the experimental setup. This is the top view of the setup showing three vacuum chambers- STM, load-lock and preparation (not to scale).

a cold ion gauge. For transferring the sample/tips from the load lock to the preparation chamber, a magnetically coupled transfer rod is used.

Most of the samples were prepared in the preparation chamber, which included cleaning of the metal substrates (Au(111), Cu(111), Ag(111)) and evaporation of different molecules. The chamber is equipped with a manipulator rod which can advance into the STM chamber by turning a crank fitted with a stepper motor. The manipulator rod of the preparation and the transfer rod of load-lock chamber have holders where the sample carrier plates can be placed. The preparation chamber has another magnetically coupled manipulator referred as the wobble stick. This wobble stick can be moved inward or outward as well in the plane perpendicular. This provides four degrees of freedom which helps to grab the sample carrier plate and to move -from or -to the manipulator rod. There is a sputter gun (Ar^+ flow) attached to the chamber for cleaning the metal crystals. A 3-cell molecular evaporator with water/LN_2 cooling (from company *Kentax*)

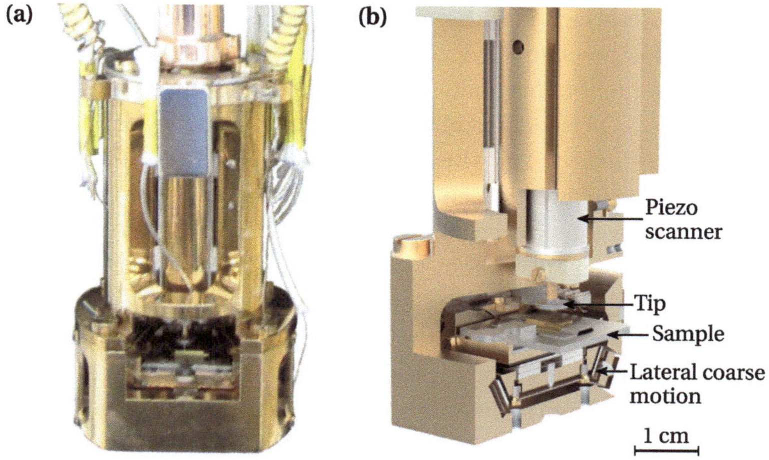

Figure 3.2: STM body. (a) Image of the just finished STM body. **(b)** A cross sectional view of the CAD drawing of the STM body showing some important parts (adapted from [74]).

is also attached to the preparation chamber, which was used for evaporating various materials such as NaCl and H_2Pc (see section 3.3.3).

The STM chamber has the LN_2 and He cryostats in which the STM is placed. This chamber also has a wobble stick. A magnifying camera is mounted, which helps to look into the STM body while placing the tip or the sample. To avoid the mechanical vibrations, the whole setup is decoupled by a pneumatic vibration isolator from company *Newport*.

3.2 STM Setup with Optical Access

The STM (referred as γ-STM) was built by Dr. Kevin Edelmann and Dr. Lukas Gerhard at *Institute of Nanotechnology*, KIT [73]. Two master students, Moritz Winkler and Lars Wilmes, also assisted during the construction of the setup. The design of the STM promises to achieve high light collection efficiency [74]. In the following subsections, a brief design of the STM, out-coupling of the light, and recording of the corresponding spectra will be discussed.

3.2.1 STM Body

Figure 3.2 shows the actual and rendered image of the STM body [73, 74]. The design of the STM is based on the Pan-style [75] and the design of Zhang *et al.* [76]. The STM body shown in the Fig. 3.2 is attached with three springs and hangs freely from the cryostat, which provides a good mechanical decoupling. The STM sits at liquid He temperature of 4.4 K which prevents the thermally induced motion of the molecules and even of tip atoms. Tip of the STM is located above the sample because the optical fiber is fed through the cryostat from the top. This way the fiber does not bend too much and any sharp bend of the fiber is avoided. Scanning with the STM tip is performed with a 30 mm long and 10 mm wide segmented lead zirconate titanate (PZT) piezo (see Fig. 3.2b) that can scan 2μm laterally and move about 150 nm vertically. For the larger lateral/vertical motion, a *coarse motion* positioning system is present in the STM (based on slip-stick motion [77]). This coarse motion can move the tip assembly vertically (12 mm) and sample laterally (8 mm $\times$ 4 mm in x $\times$ y, respectively). The whole STM body can sit on a stage (parked position) which is thermally coupled to the cryostats and can cool down the body much faster, but during the measurement, the STM body is unparked and hangs freely. For good thermal conductivity and stiffness, the STM body was constructed with beryllium copper. The STM body is accessible through two doors of the radiation shield of the cryostats. This enables to transfer the tip or sample *in-situ*. A detailed discussion of the STM design is reported in the Edelmann *et al.* [73, 74].

3.2.2 Mirror Tip

The tunnelling current and the conversion efficiency (photons/electron) in the STM junction are very low (see chapter 2), which means it is important to collect as many photons as possible. This can be achieved by covering a large solid angle around the STM junction. Typically, this can be achieved by placing lenses, mirrors, or optical fibers close to the junction [5, 7, 10, 16, 18, 63, 78–93]. Setups with lenses have to deal with the out-coupling of the emitted light to the far-field which requires focusing the lenses in the junction for every tip. Particularly, in the UHV, the geometrical focusing of the lenses is not trivial. Using several lenses, a solid angle of 27% has been achieved [82]. With mirrors, however, a coverage of 65% has be achieved [94] but they also require very precise adjustment for focusing the light onto another focal point for out-coupling. Setups with optical fibers overcome the issue of focusing the light, however, they need to be placed very close to the junction since they have a limited acceptance angle.

The γ-STM setup [73] has a parabolic mirror with an integrated tip, which overcomes the issue of focusing of the emitted light (as discussed by Khang *et al.* [87]). The focused light is then guided out of the cryostat into the spectrometer via an optical fiber, avoiding any refocusing.

Figure 3.3a shows the design (CAD) of the parabolic mirror with the integrated mirror tip. To prepare such a structure, direct laser writing (DLW) technique was utilized [95]. The mirror tip has two major components. One is the parabolic mirror and the other is

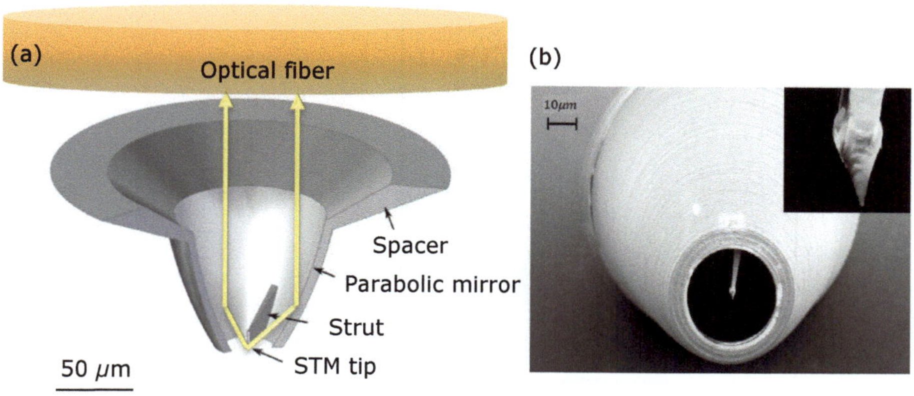

Figure 3.3: Mirror tip. (a) Cross section view of the mirror tip (CAD drawing). **(b)** SEM image of the metalized mirror tip, inset shows the the tip (adapted from [74]).

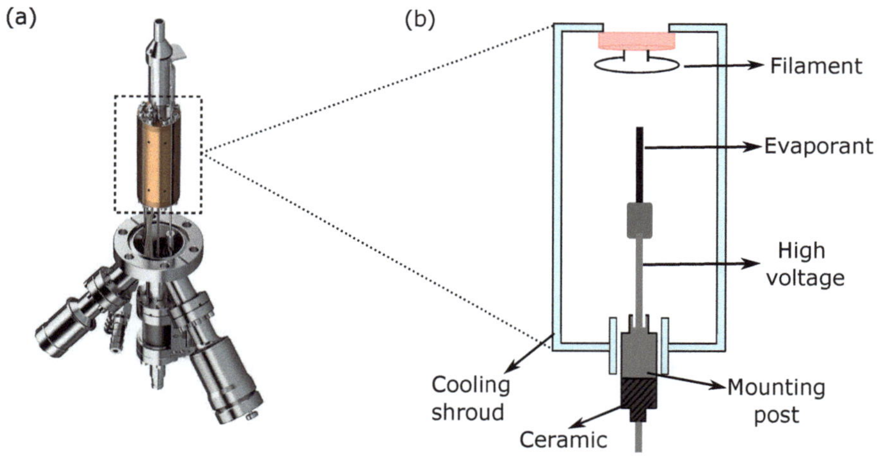

Figure 3.4: (a) Image of the electron beam evaporator from company *Focus.* **(b)** Schematic of the part marked in **a** (dotted rectangle). The evaporant (Cr and Ag) is subjected to high voltage (HV) and electrons emitted from the filaments are accelerated towards the evaporant hence increasing the temperature for the evaporation.

the tip connected to the mirror via a strut. The tip protrudes out of the mirror plane by 1.3 μm leaving a clearance angle of 5°. The apex of the tip is at the focus of the mirror. The parabolic mirror, which then collimates the emitted light from the STM junction into an optical fiber is located directly above the mirror. A SEM image of the mirror tip is shown in Fig. 3.3b. The height of the parabolic part of the mirror is 80 μm, which sits on a spacer of 30 μm, making total height of the mirror around 110 μm. The opening of the

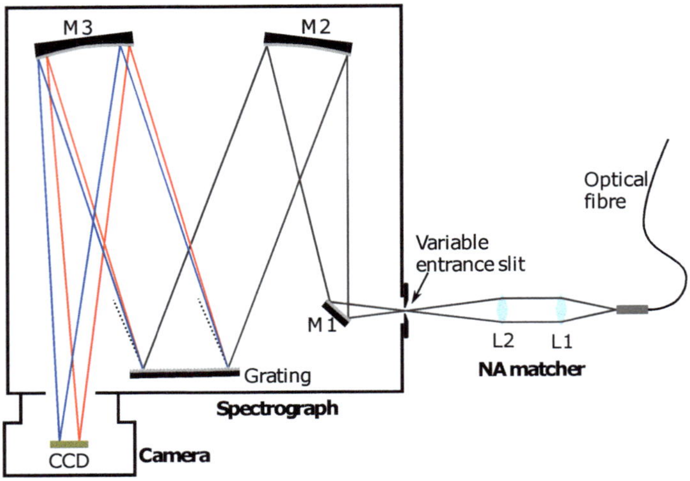

Figure 3.5: Schematic of the spectrometer and detector (adapted from [74]).

parabolic mirror on the tip side is about 30 μm in diameter whereas the other end has an opening of 100 μm. The inner base diameter of the spacer is 185 μm. The whole mirror tip structure is written on a polished steel foil around a pinhole of diameter varying between 130 and 150 μm. The optical fiber (diameter = 200 μm) comes from the other end of the pinhole, making a joint assembly of fiber, steel foil and the mirror tip.

The DLW tip is made of polymer (IP-S), which means it needs to metalize in order to make the mirror reflecting and tip conducting. To achieve that, the mirror tip was coated with 100 nm silver. As an adhesion layer, 20 nm thick chromium was used. For the metalization, electron beam evaporators (*EFM 3i*) from company *Focus* were used (see Fig. 3.4). The mirror tips are not very rigid in nature and to enhance the robustness, 20 nm thick Al_2O_3 was coated before the metalization via atomic layer deposition (ALD). Once DLW fabrication parameters were found for the mirror tip, several of them can be written on demand.

3.2.3 Spectrometer and Detector

The out-coupled light from the STM junction exits the other end of the optical fiber (fiber is $\sim$ 5.5 m long). The light exiting the other end of the fiber diverges and a lens assembly is used to first collimate and then to focus it on a commercial *Acton SP-2156* imaging spectrograph (see the schematic in Fig. 3.5). The spectrograph is equipped with a 300 grooves/mm diffraction grating combined with a *PyLoN 100B* liquid nitrogen cooled charge-coupled device (CCD) camera by *Princeton Instrument*. Three mirrors (M1, M2

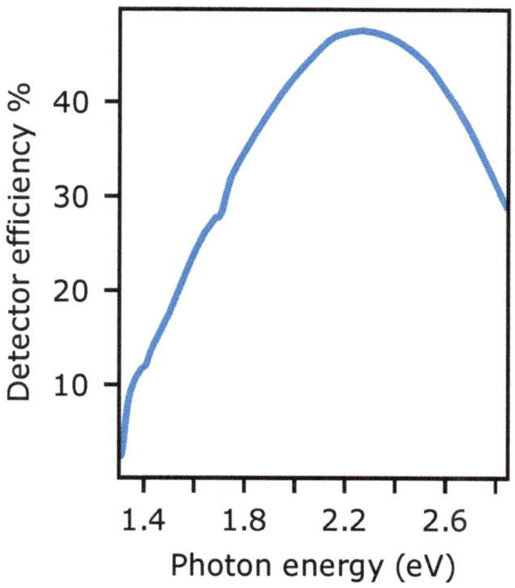

Figure 3.6: Detector efficiency of the STML setup excluding the possible losses in micro-fabricated mirror tips.

and M3) are present in the spectrograph. M1 collimates the light on to a parabolic mirror M2, which then focuses the light on to the grating. Diffracted light from the grating is then focused on the CCD by mirror M3. Some of the light intensity is lost in the mirror tip, at the fiber interfaces, lenses, and mirrors in the spectrograph. The losses in the mirror tip are unknown, however, other losses can be evaluated, including the grating efficiency and the conversion efficiency of the CCD camera. The wavelength response to these losses amounts to the detector efficiency shown in Fig. 3.6. All the spectra discussed in this thesis are corrected with the detector efficiency unless specified.

3.3 Sample Preparation

For the sample preparation, mainly two deposition methods, thermal and spray, were employed. Prior to the deposition, the metal substrate was cleaned via sputtering. In the following sections, substrate cleaning and molecular deposition methods will be discussed.

3.3.1 Substrate Cleaning

Molecules were deposited on clean Au(111) single crystal substrates. Substrates were cleaned with repeated cycles of Ar^+ ion (with 1.5 keV) sputtering and subsequent annealing at 560 °C. The sputtering removes the adsorbates whereas the annealing brings

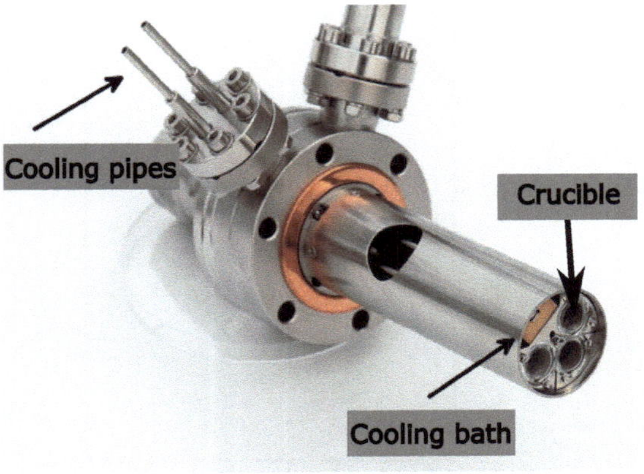

Figure 3.7: 3-cell *Kentax* evaporator with water/LN2 cooling.

impurities to the surface and also recovers defects by mobilizing the surface atoms. Some of the samples were prepared at elevated substrate temperatures. For example, NaCl bi and tri-layers were formed at substrate temperature of 450 K. To avoid undesirable heating around the sample, direct heating by electron bombardment was used. The heating was done on the manipulator head which is also equipped with a nitrogen cooled bath, which helps to quickly cool down the annealed sample. This cooling bath also helps to keep the sample temperature low if needed.

3.3.2 Thermal Deposition of Molecules

Evaporation of sublimable materials (NaCl and H_2Pc) were done by *Kentax* 3-cell evaporator equipped with cooling via water or liquid nitrogen (see Fig. 3.7). The evaporator has three separately controlled evaporation cells and the sample can be exposed to one or two cells at a time with the help of a shutter. Water cooling prohibits the unnecessary heating of the surrounding, which is important to obtain a clean sample. Each cell has a separate heating element and can be controlled via power supply.

3.3.3 Spray Deposition of Molecules

Molecules discussed in chapter 5 and 6 are not sublimable and heating them to higher temperatures results in their decomposition. To overcome this issue, spray deposition has been utilized which was developed in our group (published in Valášek *et al.*) [96]. In the published paper, however, a more sophisticated version has been reported whereas in this thesis molecules were deposited with an even simpler version of the setup. Figure

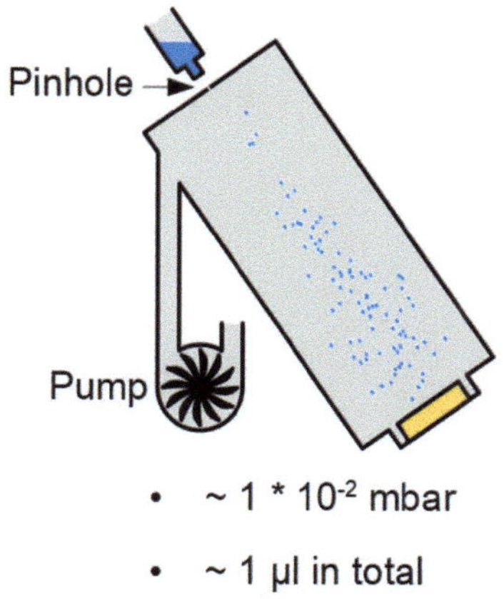

Figure 3.8: Schematic of the spray deposition setup. Initial design of the image was made by Jan Homberg.

3.8 shows the schematic of the spray setup, which includes a vacuum chamber with a pinhole attached to the load-lock chamber (see Fig. 3.1). The pressure in the chamber is $\sim 1 \times 10^{-2}$ mbar during the spray. The clean Au(111) crystal is placed directly opposite to the pin hole and then around 1 μl of molecular solution in CH_2Cl_2 is put through the pin hole. Due to the pressure difference, the solution is sucked into the chamber and small droplets land on the sample. After the spray, the sample is immediately put into the UHV condition in the load-lock chamber and then transferred to the manipulator head of preparation chamber. Thereafter, the sample is annealed at >100 °C to promote the diffusion of the molecules as well as to promote the reaction on the surface, which has been discussed in the chapter 2.

4 Boosting Light Emission from H_2Pc Molecules

Recently, STML has allowed the discovery of many intriguing phenomena at the single molecule level [5–7, 19, 21, 23–25, 29, 64, 97]. Most of the studies are focused on the underlying mechanism involved in light emission and a general concept of how to increase the overall efficiency of light emission is lacking, although it is a key factor in regard to possible applications.

In this chapter, a comprehensive study of light emission from a single Hydrogen-Phthalocyanine (H_2Pc) molecule is presented (see structure in Fig. 4.1). Most of the data presented in this chapter are published in Rai *et al.* [29]. The focus of the study is to not only understand the underlying mechanism involved in the light emission but also to propose a general pathway to actually boost the light emission efficiency of an individual molecule. The chapter will start with the sample preparation on which the STM and STML experiments were performed, followed by the discussion of the experimental results and the theoretical calculations to support them.

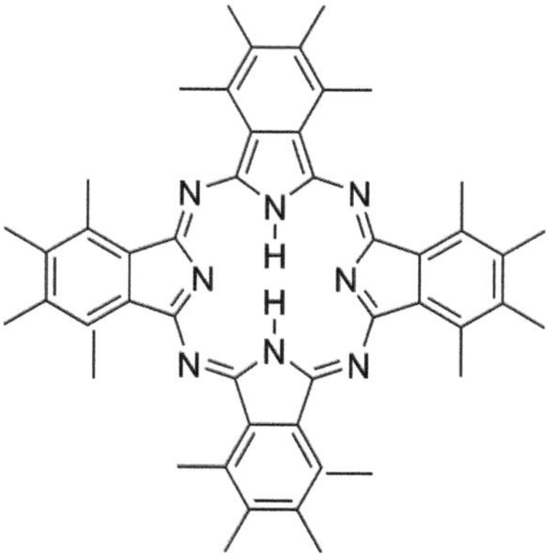

Figure 4.1: Molecular structure of Hydrogen-Phthalocyanine (H_2Pc) molecule.

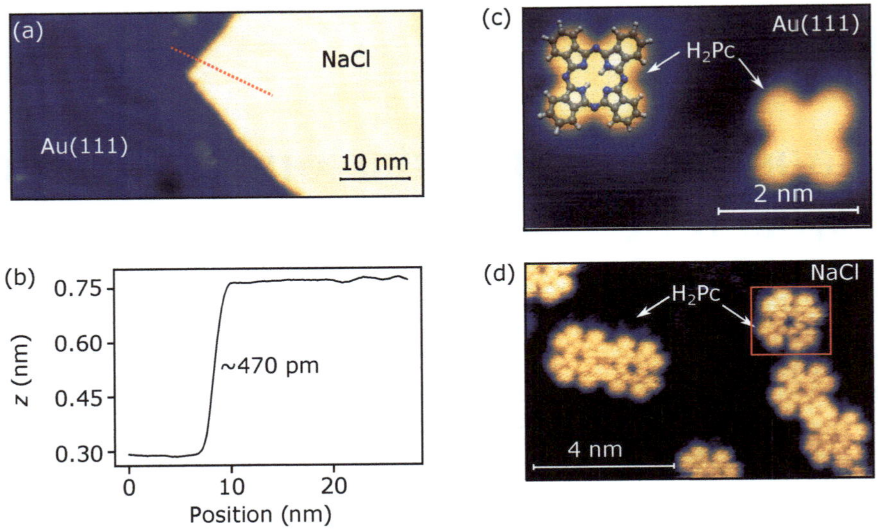

Figure 4.2: Topography of H_2Pc. **(a)** STM image of a NaCl island on Au(111) ($U = -1.6$ V, $I = 2$ pA). **(b)** Apparent height (z) of the tri-layer NaCl island on Au(111) along the line scan marked in **a** (red dotted line). **(c)** H_2Pc molecules adsorbed on Au(111) ($U = -2.4$ V, $I = 5$ pA). **(d)** H_2Pc molecules adsorbed on NaCl ($U = -1.5$ V, $I = 2$ pA). The molecule marked with a red box appears less bright compared to the others and this can be attributed to effects of molecules being adsorbed at different NaCl site than the rest (as observed in [28]).

4.1 Adsorption of H_2Pc

For STM experiments, molecules were deposited on metal surfaces to support the quantum tunnelling in the junction, however, this in turn quenches the molecular luminescence. To prevent this substrate-induced luminescence quenching, the light emitting part of the molecule needs to be electronically decoupled from the substrate [5, 7, 63, 98–102]. The most common approach to decouple the molecules from the metal substrate is by virtue of a thin insulating layer of NaCl. Here as well, tri-layer of NaCl have been used to decouple the H_2Pc molecules from Au(111) substrate.

For the sample preparation, first the Au(111) substrate was cleaned by several cycles of sputtering and subsequent annealing (see Chapter 3). NaCl was thermally evaporated on the cleaned Au(111) surface while keeping the substrate at 450 K in order to promote multi-layer formation of the salt [99]. Figure 4.2a shows the structure of a tri-layer island of NaCl on Au(111). The NaCl islands typically exhibit a square shape, having straight step edges and kinks, similar to the one shown in Fig. 4.2a. The NaCl islands do not modify the herringbone reconstruction of the Au(111) surface and the reconstruction can be observed as an additional corrugation of the islands, indicating a relatively small

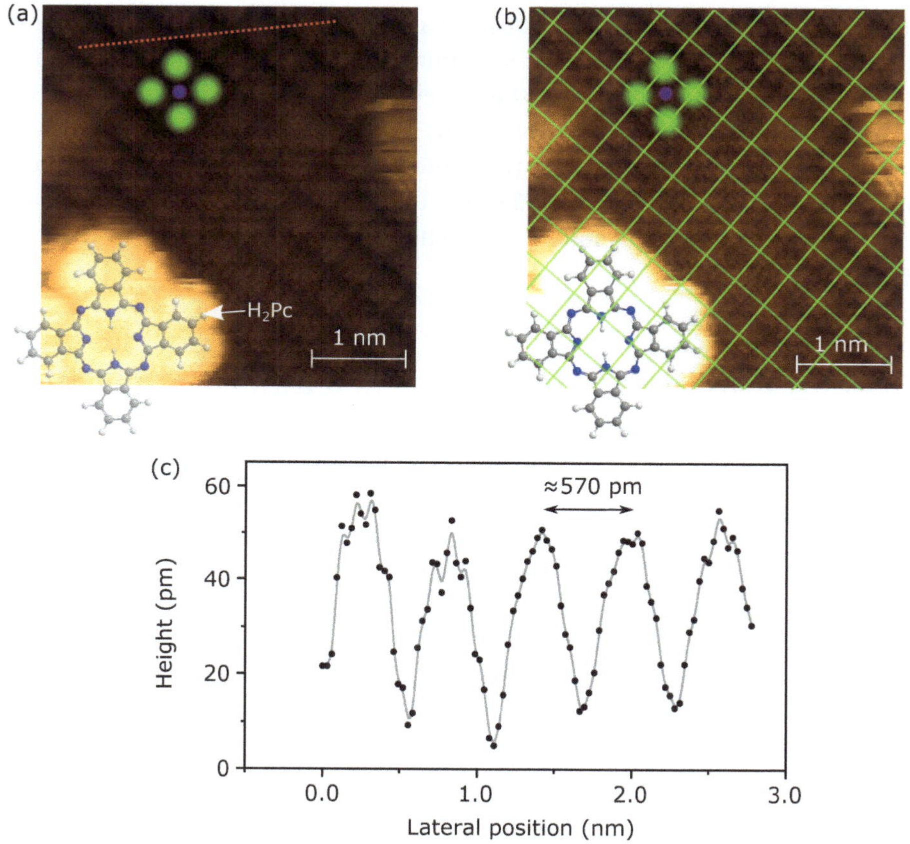

Figure 4.3: Adsorption of H_2Pc. **(a)** Atomically resolved STM image of a H_2Pc molecule on NaCl ($U = -2.7$ V, $I = 15$ pA). Estimated positions of the chlorine and sodium ions are represented with green and purple spheres, respectively. Protrusions in the NaCl islands are Cl^-. **(b)** Same image as in **a** with a grid overlayed where intersections are the estimated positions of the Cl^- and the center of the smaller squares are the positions of the Na^+. The molecular structure is superimposed on the molecule to show the adsorption configuration of the molecule with respect to the NaCl lattice. **(c)** Height profile of NaCl along the red dotted line marked in **a**.

interaction between the NaCl islands and the metal substrate. Apparent heights of the tri-layer islands are about $\sim$470 pm (Fig. 4.2b), which is consistent with previous reports [99, 103]. The subsequent evaporation of H_2Pc was performed at a substrate temperature of ~ 50 K to reduce the mobility of the molecules on NaCl.

H_2Pc molecules adsorbed on Au(111) show a four-lobe pattern in STM images (see Fig. 4.2c) which agrees with the molecular structure (superimposed). Molecules on NaCl are electronically decoupled from the substrate and the broadening of the orbitals due to the influence of substrate is prevented. Hence the molecules show an even more detailed structure (see Fig. 4.2d). Figure 4.3a shows an atomically resolved image of a H_2Pc molecule adsorbed on NaCl. Typically, such images are acquired with CO functionalized tip [18, 104, 105], however, it can also be achieved by a Ag tip as shown in Fig. 4.3a and b. In this (001)-terminated NaCl, protrusions are attributed to the Cl^-. From the high resolution image it can be said that the center of the H_2Pc is on top of the Na^+ site, consistent with previous reports [18, 28, 105, 106]. The direction of a molecular axis of H_2Pc is aligned with the [100] and [010] directions, similar to the report by Imada *et al.* [106].

4.2 Light Emission from Uncharged and Charged H_2Pc

Figures 4.4a and b show the schematic of the experimental STML setup and a typical optical emission spectrum acquired at negative sample bias. The typical quantum efficiency of the system is in the order of $\sim 10^{-5}$ photons/electron [3]. The spectrum shows three peaks (X_x^+, Q_x and Q_y), of which Q_x and Q_y can be assigned to the well-known transition of the uncharged molecule from the lowest lying singlet states of H_2Pc to the ground state [18, 21, 24, 107–110]. In contrast, the very bright and sharp ($\sim$11 meV) X_x^+ peak at $\sim$1.38 eV is newly found (Fig. 4.4b) [29]. In a recent STML study on ZnPc on NaCl/Au(111), an additional peak at similar energy has been observed, and it was assigned to the positively charged molecule [23].

For a better understanding of the observed spectral lines of H_2Pc, quantum chemical simulations of the uncharged H_2Pc and charged H_2Pc^+ molecules were performed [111–113]. All theoretical calculations were performed by the group of Prof. Dr. Carsten Rockstuhl at *Institute of Theoretical Solid State Physics*, KIT. The gas phase spectra of H_2Pc (black line) and H_2Pc^+ (red line) in Fig. 4.4c reveal that the X_x^+ excitation indeed stems from the charged molecule. Furthermore, the peak is split into two components (X_x^+, X_y^+), similar to the Q_x and Q_y lines of the uncharged molecule. Figure 4.4d shows the natural transition orbitals (NTOs) of the two calculated X^+ excitations of the charged system. The NTOs reveal the character of the observed X^+ transitions, assigning them to a transition from the highest lying occupied state to the lowest lying unoccupied state of the charged H_2Pc^+ molecule. Due to the Coulomb interaction, the energetic position of the lines slightly shifts by the presence of nearby charge (blue line in Fig. 4.4c), which is used to reflect the presence of the tip close to the molecule. The calculation also reproduces the two experimentally observed transition energies of the uncharged molecule. In the experiment, the higher energy line Q_y is, however, less intense. This was attributed to the fast internal conversion of the excited Q_x and Q_y states [114]. Similarly, the higher energy line of the charged molecule X_y^+ is not observed in the experiment on the level of the background intensity. Due to the high cost of calculation, the exact

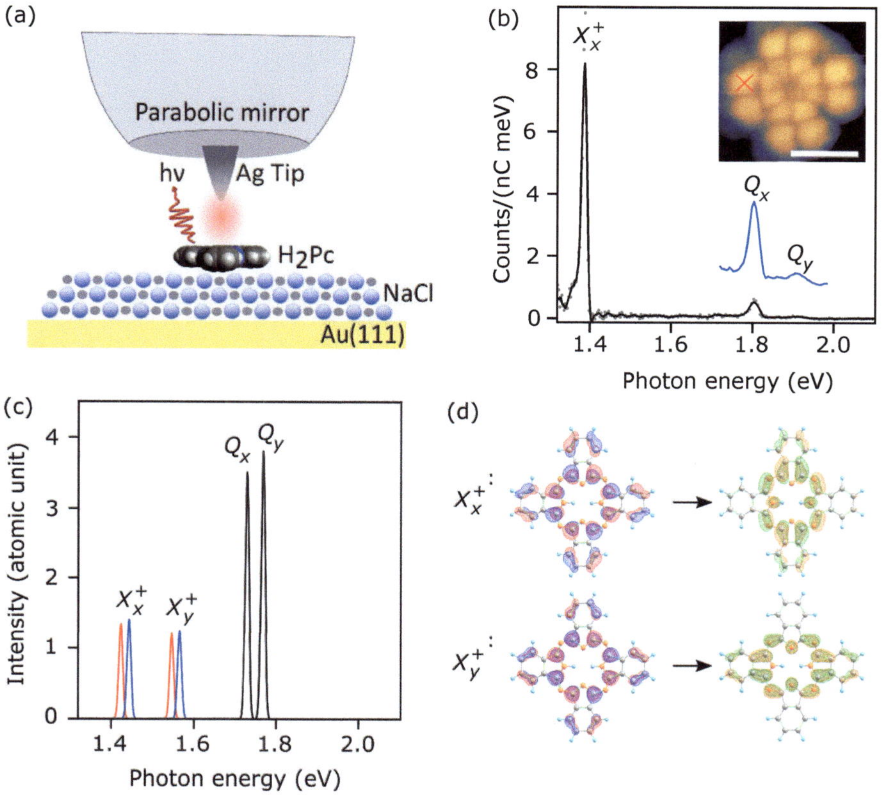

Figure 4.4: STML and calculated emission lines of H_2Pc. (a) Schematic of the experimental setup including a 3D printed parabolic tip (not to scale) [73]. The single H_2Pc molecules are decoupled from the Au(111) surface by three layers of NaCl. **(b)** In black, the optical emission spectrum recorded with the tip positioned at the edge of a H_2Pc molecule (red cross in the inset topography, $U = -2.2\,\mathrm{V}$ and $I = 2\,\mathrm{pA}$) at negative sample bias ($U = -2.7\,\mathrm{V}$, $I = 10\,\mathrm{pA}$, $t = 20\,\mathrm{s}$). Spectrum is replotted to highlight the intensity of Q_x and Q_y peaks (in blue) with five times scale. Length of the scale bar of the inset topography is $1\,\mathrm{nm}$. **(c)** Calculated photoabsorption spectra of the uncharged (Q_x and Q_y) and the charged (X_x^+ and X_y^+) molecule without (red line) and with (blue line) a negative point charge placed at the tip position. **(d)** Orbitals involved in the optical transitions of the charged system. The natural transition orbitals (NTOs) (hole: red/blue, particle:green/orange) of the corresponding excitations. Theoretical calculations were performed by our collaborator Dr. Christof Holzer at *Institute of Theoretical Solid State Physics*, (KIT).

geometry of the junction is not taken in the calculations and hence the intensities of the peaks cannot be compared.

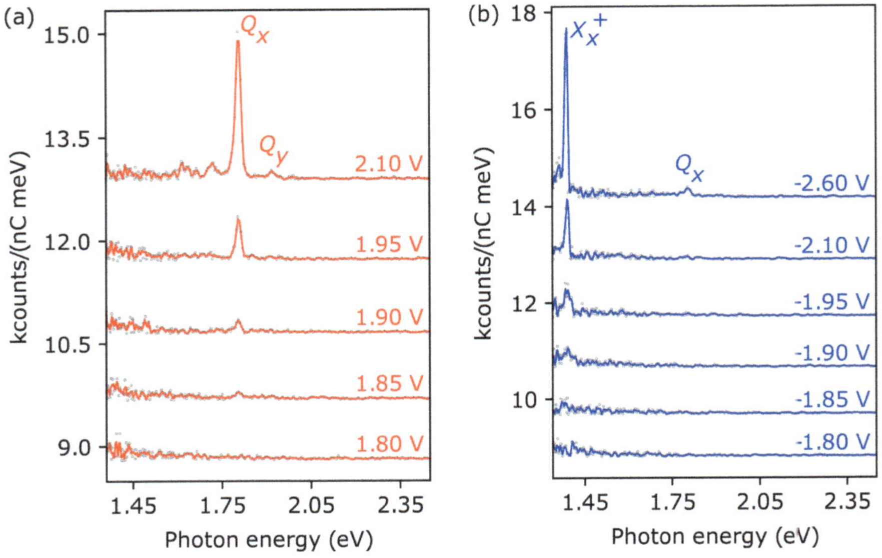

Figure 4.5: Energetics of the emission of the Q_x and X_x^+ lines. (a)-(b) Optical emission spectra at different sample bias voltages ranging from 2.10 V to −2.60 V (I = 25 pA, t = 5 s).

4.2.1 Energetics

For the further understanding of the nature of the new X_x^+ spectral line, the evolution of optical emission spectra at both bias voltage polarities were measured, see Fig. 4.5a and b. In contradiction to the previous reports, emission of Q_x line is observed for both polarities. Also, as soon as the energy of the exciting electron is equal to the energy of the $Q_x(Q_y)$ line, which is $\sim$1.80 eV($\sim$1.92 eV), emission is observed into the far-field. These observations suggest that the emission of $Q_x(Q_y)$ line is driven by the inelastic energy transfer (IET) via plasmon [23] (see chapter 2). The bias-dependent measurements also show a phenomenon called up-conversion electroluminescence (UCEL) for $|U| \leq 1.8$ V with low count rates (see Fig. 4.6a) similar to the observation made by Chen *et al.* [24]. Possible mechanisms for such emission have been discussed in chapter 2.

Figure 4.5a and b show that the intensity of the Q_x line is significantly lower, which is a consequence of dipole moment of the involved molecular excited states being oriented in-plane, this suppresses the emission to the far-field [18, 21, 29].

Furthermore, X_x^+ is found only at negative sample bias in agreement with the charged state of the molecule. Emission of X_x^+ sets in at −1.8 V (Fig. 4.5b and 4.6b). To understand this onset of X_x^+ peak, differential conductance (dI/dU) measurement was carried out on the adsorbed molecule. Figure 4.6b shows the integrated photon count of the line X_x^+ as a function of U together with the dI/dU. dI/dU shows two peaks

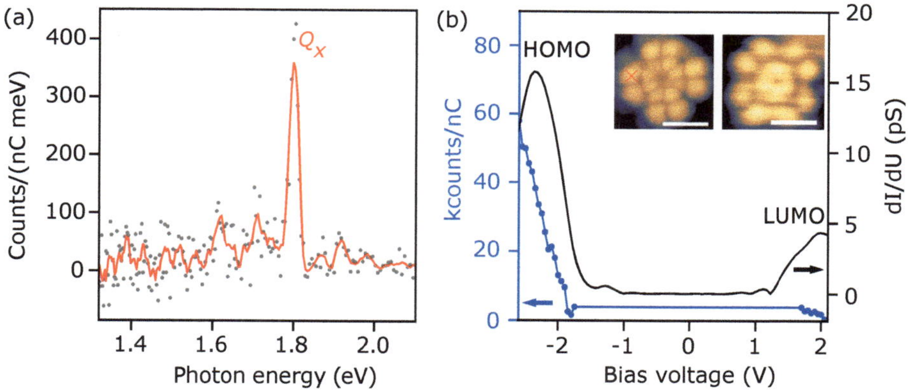

Figure 4.6: Energetics of the emission lines of H_2Pc. (a) Optical emission spectrum showing UCEL (U = 1.6 V, I = 3.1 pA, t = 10 s). **(b)** dI/dU spectrum (black curve) recorded on a single H_2Pc molecule and photon count integrated over the X_x^+ peak (1.373 eV - 1.401 eV) as a function of sample bias (blue dots and line). Insets show the structure of HOMO (left, scanned at $U = -2.2$ V) and LUMO (right, scanned at $U = 2.3$ V). Scale bar is equal to 1 nm.

at -2.3 V and 1.7 V which can be assigned to the highest occupied molecular orbital (HOMO) and the lowest unoccupied molecular orbital (LUMO), respectively. In STM images, HOMO and LUMO can be imaged by scanning the molecules in the corresponding energy window [18, 106]. The inset of Fig. 4.6b shows the HOMO (right image with four-fold symmetry), and the LUMO (left image), similar to previous observations [18, 24, 28, 106]. The energetic onset of the X_x^+ line coincides with the HOMO peak, which means electrons may tunnel elastically from the HOMO to the empty states of the tip, hence charging the molecule.

4.2.2 Underlying Mechanisms

Mechanisms to explain the emission of Q_x and X_x^+ lines are depicted in Fig. 4.7a and b. Here, based on our observations, the mechanism that explains the origin of uncharged Q_x peak is based on IET via plasmon (see Fig. 4.7) and discussed in all detail in chapter 2.

The mechanism that explains the emission from the charged molecule is depicted in Fig. 4.7b. At sufficiently high bias ($U < -1.8$ V), the Fermi energy of the tip is in resonance with HOMO of the molecule, allowing an electron to tunnel from the molecule to the tip. This leaves the molecule in a charged state. This charged molecule may then be stable long enough so that the energy lost by the an inelastically tunnelling electron can be transferred to this charged molecule that undergoes an excitation-emission cycle. The

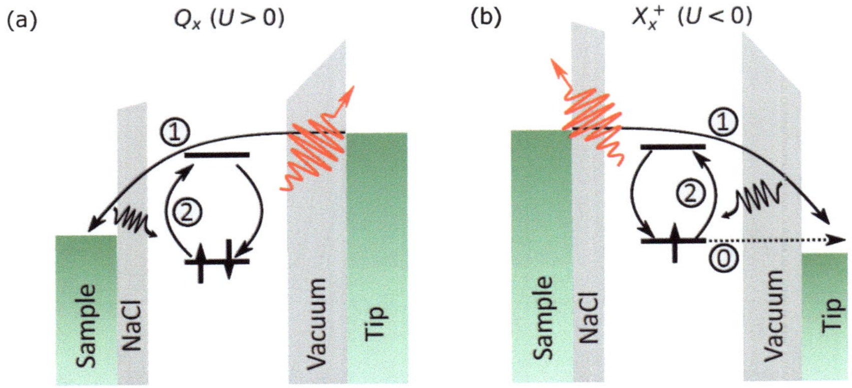

Figure 4.7: Underlying mechanisms. **(a)** Plasmon is excited by an inelastically tunnelling electron (process 1). The plasmon (black wavy arrow), in turn, excites the molecule (process 2), which decays under reemission of a photon (red wavy arrow). **(b)** Emission of the X_x^+ line requires charging the molecule by elastic tunnelling of an electron from the HOMO to the tip (dotted arrow, process 0) before the excitation-emission cycle.

cationic form of the molecule can be generated in the negative bias only, which explains the fact that X_x^+ line is not observed at positive bias, because at positive bias the voltage drop would be across the vacuum between the tip and the molecule.

Previous reports [5, 16] suggest that the high density of plasmonic modes enhances the emission efficiency of the emitter, *i. e.* if the plasmon is located at the Q_x or X_x^+ peaks, their respective emissions will be amplified. This is a consequence of Purcell effect [61]. For this reason it is important to mention here that the plasmon resonance of the Ag tip used throughout this work is centred around 1.8 eV, see Fig. 4.8b (detector efficiency is plotted in Fig. 4.8a). Hence, the much higher intensity of the X_x^+ line in comparison to the uncharged emission hints for a more efficient coupling of the charged molecule to the electric field and is not due to the plasmonic resonance. This will be discussed in more detail in the section 4.2.5. Also, both emission lines, Q_x and X_x^+ are present in the negative bias, which indicates faster transitions between the differently charged states than our recording time (1 s).

As far as the uniqueness of the depicted model is concerned, it can be said that the proposed model describes the observed phenomenon of light emission from the charged species at negative sample bias and from the uncharged molecule at both bias polarities. Other models have also been proposed to capture other aspects [24, 25], but for the observations discussed here, model in Fig. 4.7 is sufficient.

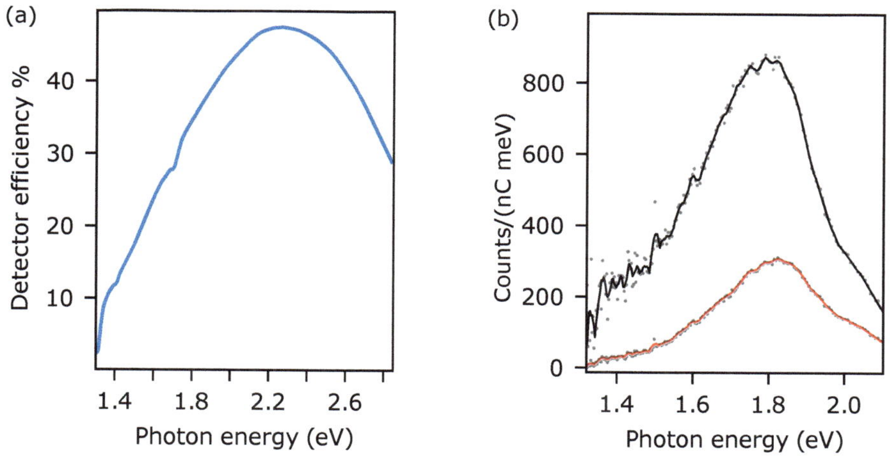

Figure 4.8: Detector efficiency and plasmon response (a) Detector efficiency of the light collection setup excluding the microfabricated mirror tip. **(b)**, Plasmonic light emission of the mirror tip used throughout this work. Red/black lines show the signal uncorrected/corrected for detector efficiency.

4.2.3 Spatial Variation

As per the model in Fig. 4.7b, X_x^+ emission happens once the HOMO of the molecule comes in resonance with the Fermi energy of the tip. In agreement with the model, the spatial distribution of the intensity of the X_x^+ line reflects the HOMO distribution. This is illustrated in Figs. 4.9a and b which show the topography of a single H_2Pc molecule and the integrated intensity of the emission at the X_x^+ line (1.373 eV-1.401 eV) on a grid of 36×33 pixels [7, 22, 101, 115]. Thus, the charging probability of the molecule as the initial step of the emission process is directly related to the energy and shape of the HOMO [116] as is shown in Fig. 4.9c and it enters the intensity of light emission. However, the inner lobes of the HOMO do not appear bright and the reason for this will be discussed in the following.

The absence of light emission with the tip placed in the center of the molecule can be understood in terms of variation of the out-of-plane component of the oscillator strength. For this purpose, time-dependent density functional theory (DFT) calculations are performed by Christof Holzer. Time-dependent DFT simulations predict that the out-of-plane (z-component) of the dipole moment (p_z) to be vanishingly small if a point charge, simulating the tip is placed at the center of the molecule (Fig. 4.9a). Moving the point charge *i.e.* the tip outwards then gives rise to a non-vanishing p_z as shown in Fig. 4.10a. This emergence of p_z in turn boosts the light emissions into the far-field which

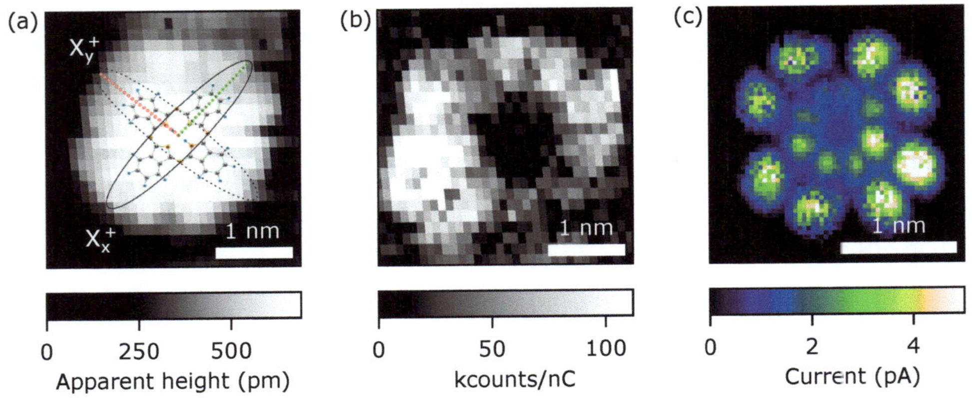

Figure 4.9: Spatial variation of the emission of the X_x^+ line. (a) Topography of the H_2Pc molecule during the recording of the photon map with molecular structure superimposed to scale. Solid (dashed) ovals indicate the orientation of the orbitals involved in light emission of X_x^+ (X_y^+) lines. **(b)** Spatial distribution of integrated photon count in the energy range of the X_x^+ line (1.373 eV - 1.401 eV, $U = -2.5$ V, $I = 2$ pA, $t = 10$s). **(c)** Current map of the H_2Pc molecule recorded at constant height at $U = -2.3$ V representing the HOMO of the molecule.

can be confirmed by plotting the radial dependence of the measured photon count, see Fig. 4.10b. A maximum of the calculated intensity is found when the point charge is moved $\sim$0.5 nm outside the center, which is in good agreement with the experimental findings.

The qualitative behaviour of Fig. 4.10a and b agrees well, however, the experimental curve is wider. This can be explained by the real tip being less sharp than an ideal point charge. p_z^2 is still low at the positions of the inner lobes of the HOMO, explaining the absence of these lobes in the experimental photon map as well.

It is very interesting to see that when moving the tip perpendicular to the hydrogen bond direction (green dots in Fig. 4.9a), the X_x^+ line becomes bright and the X_y^+ line vanishes whereas when going along the hydrogen bond direction (red dots in Fig. 4.9a), the situation reverses. In the experiment, however, the X_y^+ line is suppressed due to internal conversion as discussed in Baeten *et al.* [114]. In the mentioned report, it was observed that the Q sub-bands of H_2Pc (Q_x and Q_y) have a small energy gap and the fast interconversion rate, at which the equilibrium is established. Excitation to the Q_x state will result in emission not only from the Q_x state, but also from the Q_y. Due to the low population of the more energetic Q_y state, the intensity of the corresponding emission is small compared with that originating from the Q_x state.

Following the above argument, the expected photon map should only show the emission from the four lobs perpendicular to the hydrogen atom (along the green line in Fig. 4.9a). However, we observed light from all eight lobs of the HOMO. This can be understood

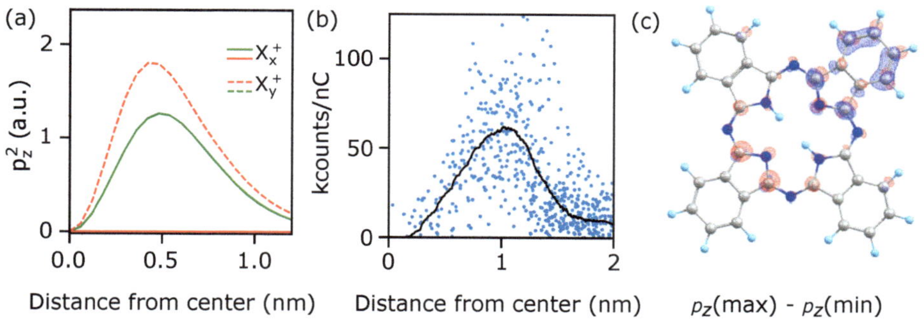

Figure 4.10: Variation in emission probability (a) Green (red) lines depict the squared simulated out-of-plane component of dipole moment p_z^2 for a point charge (tip) located at the positions indicated by green (red) dots in Fig. 4.9a. Solid (dashed) lines represent the X_x^+ (X_y^+) emission lines. The red solid line lies above the green dashed line. **(b)** Integrated photon count of the X_x^+ line for each pixel of the photon map (blue dots) and smoothed over 197 points (black line) as a function of distance from the centre of the molecule. **(c)** Difference of density of HOMO with tip at $r = 0.0$ and at the position with maximum z-component.

by the fact that the H_2Pc molecule has fast transition of the hydrogen atom between the two neighbouring equilibrium sites, *i.e.* tautomers of H_2Pc. A detailed discussion is presented in the section 4.2.4. Furthermore, difference in HOMO with tip at $r = 0.0$ nm and at the position with maximum z-component is finite.

4.2.4 Identifying the Tautomers

Figure 4.11a shows the spatial distribution of X_x^+ peak over the same grid as the photon map (see Fig. 4.9b). This distribution map of peak positions shows four segments with two distinct peak positions, a low energy peak (red area) and a high energy peak (blue area). Figure 4.11b shows the two spectral lines averaged over the red and blue areas in Fig. 4.11a which are split by about ~ 10 meV. Understanding of this behaviour can be realized by looking at the structure of the H_2Pc molecule (see Fig. 4.11b). H_2Pc has two hydrogen atoms in the center, bonded to two nitrogen atoms. There are two further energetically degenerate sites available for the hydrogen atoms. For this reason, the hydrogen atoms may switch between the two neighbouring nitrogen atoms forming two configurations of same energy, given that the required barrier can be overcome. This phenomenon is called tautomerization. This has been observed in other STM as well as a STML studies [28, 117]. In STML, this has been observed for the neutral H_2Pc in Doppagne *et al.* [28]. The emissions from two different tautomers of the charged H_2Pc molecule with slightly different energies due to a different interaction with the substrate. For this, the local 4-fold symmetry of the NaCl layer in combination with the molecule needs to be broken. This may arise from the interaction with the Au(111)

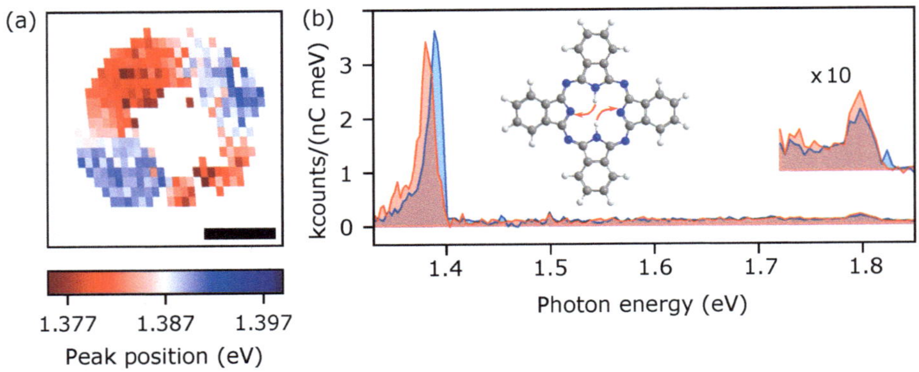

Figure 4.11: Tautomerization of H_2Pc (a) The spatial distribution of the peak positions of X_x^+ spectral line shows four segments with two distinct peak positions, a low energy peak (red area) and a high energy peak (blue area). This is the extracted from the photon map shown in Fig. 4.9. **(b)** Spectral lines averaged over the red and blue areas in **a**. At higher energies, the spectrum is plotted with 10 times the scale to highlight that the Q_x peak. It has very low intensity and does not reflect the tautomeric behaviour as of X_x^+. The molecular structure shows the tautomerization by mapping of the hydrogen atoms (red arrows).

surface showing only 3-fold symmetry. The tautomerization of the charged neutral molecule was found to be triggered by inelastically tunnelling electrons and was on a much faster time scale than the recording time of the optical spectra [28, 117–122]. If this applies also to the charged molecule, light emission would be observed from the X_x^+ spectral line from both of the tautomers and thus on all eight lobes of the molecule. This answers the question raised in the section before (section 4.2.3).

4.2.5 Current Dependent Measurements

To quantify the difference in the quantum efficiencies of the Q_x and X_x^+ lines, their intensities were measured as a function of the tunnelling current I at a fixed sample bias of -2.5 V (Fig. 4.12a). As the current is increased, the intensities of both Q_x and X_x^+ lines increases, but with different rates. The intensity of the Q_x line scales linearly with current suggesting a one electron process (orange dashed line in Fig. 4.12a), while the X_x^+ intensity has a quadratic dependency indicating a two electron process (black dashed line in Fig. 4.12a), which is in agreement with the model. As charging is driven by electron tunnelling from the molecule to the tip across the vacuum barrier and uncharging by electron tunnelling from the sample across the NaCl layer, the change of the tunnelling current (vertical movement of the tip) mainly influences the charging probability and leaves the uncharging probability untouched. Increasing I thus shifts the balance of the molecule to the charged state. As the molecule can either be charged or

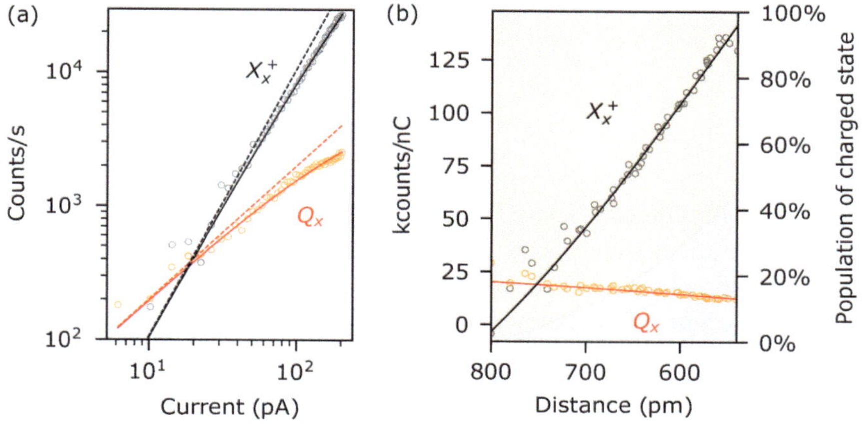

Figure 4.12: Current dependence of Q_x and X_x^+ lines. (a) Intensities of the X_x^+ (black dots) and Q_x (orange dots) lines as a function of the tunnelling current at $U = -2.5$ V. Black and red solid lines are fits according to the rate equations 4.4 and 4.5. Dotted lines are power law approximations. **(b)** Same data as in (a) normalised by tunnelling currents as a function of the distance between tip and molecule. The black (red) background depicts the population of the charged (uncharged) state.

uncharged, the two emission lines compete. This can qualitatively explain the observed deviations from the simple power laws, i.e., with the increase of the charged population, the increase of the Q_x line becomes weaker.

The above described process can be modelled using rate equations assuming that the excitation into the charged state is linear with current (αI), while the de-excitation rate (γ) back to the uncharged state is constant since the tunnelling current only affects the charging probability (modelling was done by Dr. Taavi Räpen at *Institute of Nanotechnology*, KIT. Rate equations for the charged and uncharged states can be written as:

$$\frac{dp_u}{dt} = -\alpha I p_u + \gamma p_c \tag{4.1}$$

$$\frac{dp_c}{dt} = \alpha I p_u - \gamma p_c \tag{4.2}$$

where αI is the charging rate, p_u is the populations for uncharged state and p_c is for charged states. Solutions to the equations are:

$$p_u = \frac{\gamma}{\gamma + \alpha I} \tag{4.3}$$

$$p_c = \frac{\alpha I}{\gamma + \alpha I} . \tag{4.4}$$

The measured light intensities can then be expressed by $I_u = \eta_u p_u I$ and $I_c = \eta_c p_c I$, where the "radiation efficiencies" η_u, η_c contain in themselves all physical processes from the excitation of a photon to the measurement in the detector. We assume that the radiation efficiencies are independent of the current (or equivalently vertical tip position). Fitting of the above equations to the uncharged and charged lines leads to excellent agreement (cf. solid lines in Fig. 4.12a).

To obtain the efficiency of charge-to-photon conversion, the photon rates were normalised by the tunnelling current and are plotted as a function of relative z position of the tip (see Figure 4.12b) estimated from Simmons rule[123]. If the radiation efficiencies of charged and uncharged species were equal, the intensity of the charged species X_x^+ (black dots in Fig. 4.12b) should increase (ΔI_c) by the same amount as the intensity of the uncharged species Q_x (orange dots in Fig. 4.12b) decreases (ΔI_u). In contrast, the increase of the intensity of X_x^+ clearly exceeds the decrease of Q_x. The fit for the two spectral lines gives a ratio of $\eta_c / \eta_u = 18.8$. In other words, it can be deduced that the changes in normalized intensities (ΔI_c, ΔI_u) when the tunneling parameters are varied and thus the change in the populations Δp_u and Δp_c. As the molecule is either charged or uncharged, then $\Delta p_u = -\Delta p_c$ and any difference in the change in intensity is related to η_c and η_u: $\frac{\Delta I_c}{\Delta I_u} = -\frac{\eta_c}{\eta_u}$ and ≈ 19 times more photons per tunnelling electron from the charged molecule are detected than from the uncharged molecule.

For clarity, the population of the charged and uncharged state is also indicated in Fig. 4.12b. There are many aspects involved in the high light emission efficiency of the charged molecule but the leading contribution can be assigned to the additional vertical dipole moment.

The effect of the emerging vertical dipole moment can also be realized by calculating the antenna efficiency. The antenna efficiency of the system is the fraction of the overall extracted energy from the emitter radiated into the far-field. In the low pumping regime, the overall amount of emission is proportional to the pump rate (current) and the antenna efficiency, the remaining are non-radiative transitions. The full-wave EM simulations show that the charged molecule has lower efficiency than the uncharged, which is in contradiction with the experimental observation (Fig. 4.13). However, quantum chemical calculations suggest that the transition dipole moment of the charged molecule is no longer parallel to the substrate. Including the small vertical dipole moment compared to the parallel dipole ($p_z / p_x = 0.02$) into the full-wave calculations (in

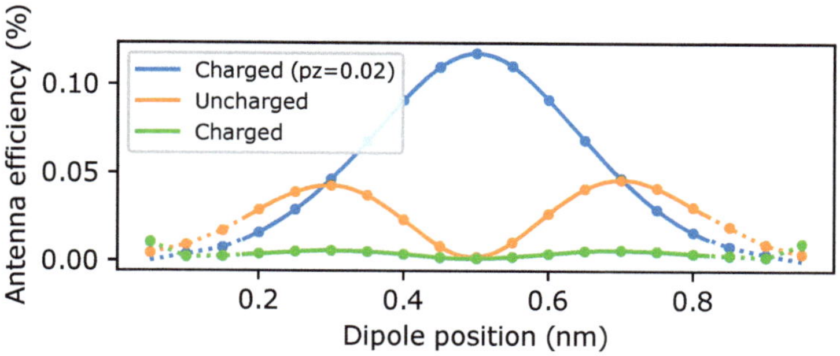

Figure 4.13: Antenna efficiency of charged and uncharged molecules. Antenna efficiency of the uncharged molecule (orange curve) together with charged molecule with and without perpendicular dipole moment, in green and blue curve. Dotted line shows the points where results do not converge.

the same geometry as before), emission into the far-field is enhanced compared to the uncharged molecule Fig. 4.13.

4.3 Summary

The focus of this chapter was to study the processes involved in the efficient light emission from a single molecule and to develop a general concept for future design of molecules to have enhanced emissions in such junctions. For this purpose, a simple system of H_2Pc molecule decoupled with thin NaCl layers from the Au(111) surface was used.

Optical emission spectra of H_2Pc molecule show the well-known Q band peaks, Q_x and Q_y. Contrary to the previous reports, a new sharp peak (X_x^+) is observed at low energy ($\sim$1.39 eV) with higher intensity than the Q_x and Q_y peaks. This newly found peak does not corresponds to any known transition of neutral H_2Pc molecule. To understand the origin of this peak, quantum chemical calculations were performed which clearly show that the X_x^+ peak is actually a transition of cationic from of the molecule (H_2Pc). Furthermore, the evolution of X_x^+ with applied bias voltage was measured which show that the X_x^+ peak is only observed at negative bias voltage. This observation hints towards the fact that this emission line is from the charged species of the molecule. To charge the molecule, an electron is removed from the molecule, which means that the emission will strongly depend on the HOMO distribution of the molecule. This reflects in the differential conductance measurement where the onset of the X_x^+ peak coincides with the HOMO peak, i.e., indicates that for light emission, an electron has to be removed from the HOMO. As a consequence, the intensity of X_x^+ peak shows same spatial distribution (photon map) as the HOMO of the molecule.

Emission of the charged and uncharged H_2Pc can be explained by the model of IET via plasmon proposed by Doppagne *et al.* [23]. Moreover, the distribution of the peak positions of the X_x^+ shows a phenomenon called tautomerization. It is remarkable that the H_2Pc molecule in its charged state also shows this behaviour which has only been reported before for the uncharged H_2Pc in such STML measurements [28].

The emission spectra clearly show that the intensity of X_x^+ is higher than the Q band peaks. To quantify the emission efficiencies of X_x^+ and Q_x peaks, current dependent measurements were performed. Using a simple population model of the molecule in two states (charged and uncharged), it was shown that the emission of the charged species is $\approx$19 times more efficient than the uncharged species. To explain this boost in efficiency, time-dependent density functional theory calculations were performed where a point charge (mimics the tip) is moved from the center of the molecule to outwards, the transition dipole of the molecule is no longer have a purely in-plane component but also has a finite perpendicular component. From this, it can be said that the perpendicular component of the dipole facilitates more efficient out-coupling of the local excitation to the far field.

This study provides a general guideline for the future development of light emitting molecular sources, which essentially require them to have a large vertical component of the molecular dipole moment. Findings of this study are used to achieve light emission from molecules which are directly deposited on the metal surface and will be discussed in the following two chapters.

5 First Steps Towards Self-decoupling in Tpd-hNDI, Tpd-sNDI and Tpd-nNDI Molecules

In the previous chapter, it was discussed that the electronic decoupling of a molecule from the metal substrate can be achieved by inserting a thin insulating spacer (organic/inorganic) between the molecule and the metal substrate. This way, fluorescence quenching caused by direct electron transfer between the molecule and the substrate is prevented [5, 63, 98, 99, 124]. A promising alternative is to decouple the molecule by functionalizing the light-emitting chromophore to a suitable spacer that preferentially binds to the metal and decouples the chromophore. Such kind of self-decoupled molecules can then be deposited directly on a metallic substrate. This allows to study wide range of chromophores. This approach is less explored due to the complex adsorption configuration of the molecules. Nevertheless, this chemical route has been tried in the past with some success, but the majority of the molecules lie flat on the metal surface and suffer severe fluorescence quenching [14–17]. To overcome this issue, a new chemical design was proposed in Balzer *et al.* [31]. In this design, a light emitting chromophore is attached to a tripodal platform which acts as a spacer and provides the desired electronic decoupling of the chromophore. Based on this design, three molecules, Tpd-hNDI, Tpd-nNDI and Tpd-sNDI were synthesised by our collaborators from the group of Prof. Marcel Mayor at *Institute of Nanotechnology*, (KIT) (see Fig. 5.1). The tripodal platform (-Tpd) is connected to the chromophores (-hNDI, -nNDI, -sNDI) by a phenylacetylene linker.

In this chapter, a detailed STM and STML study of these three molecules is presented. Some of the results discussed here are already published in Balzer *et al.* [31]. This chapter is structured in the following manner: First, we give a brief discussion of molecular structure, followed by the sample preparation. Then, the STM/STML measurements will be discussed in all details to understand the adsorption configurations of the molecules. Finally, the light emission from Tpd-sNDI molecules will be discussed, where it was observed that the Tpd-sNDI can be switched into a luminescent state with the help of the STM tip. Experimental results were then compared to quantum chemical calculations to understand the origin of the light emission from such molecules.

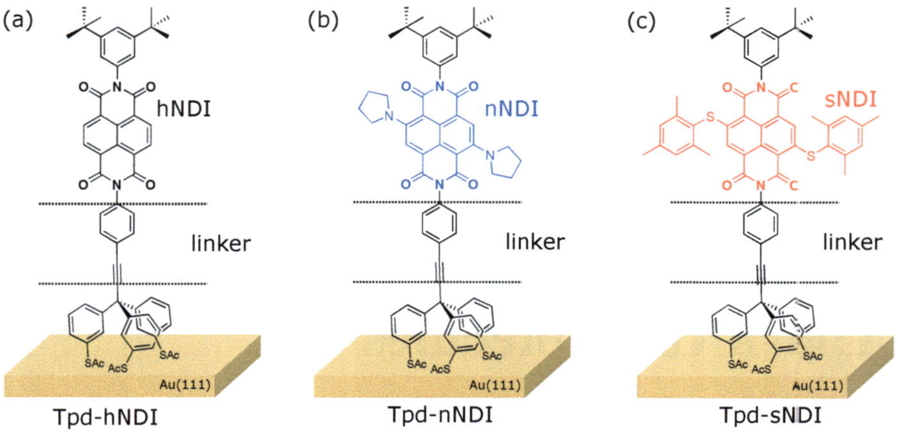

Figure 5.1: Molecular structure. (a-c) Chemical structures of Tpd-hNDI, Tpd-nNDI and Tpd-sNDI, constituting the tripodal platform connected with phenylacetylene linker.

5.1 Adsorption Configurations and Luminescence Properties

Figure 5.1a-c show the chemical structures of Tpd-hNDI, Tpd-nNDI and Tpd-sNDI in the expected adsorption configuration on Au(111). All three molecules were deposited onto a clean Au(111) surface via a spray method described in chapter 3 (section 3.3.3) [31, 96]. For the spray, $\sim 1\mu L$ solution ($\sim$1 mg/mL) of molecules in CH_2Cl_2 were used. After the spray, the samples were annealed at different temperatures in UHV to promote the deprotection of the sulfur anchor groups, which is expected to lead proper binding to the gold substrate. Moreover, annealing can also lead to the formation of ordered islands of the molecules [125–129]. In the following sections, the adsorption configurations of the three molecules and their light emitting properties will be discussed.

5.1.1 Tpd-hNDI Molecule

After the deposition of Tpd-hNDI, the sample was annealed in UHV at 100 °C. Figure 5.2a shows a typical topography. The islands seem to be composed of identical objects with clear substructures. However, the lack of any long-range order makes an assignment of the molecular adsorption nontrivial. The height profile of such an island is shown in Fig. 5.2b. The apparent height of the island is about 0.32 nm which indicates that the molecules are not adsorbed as intended but rather adsorb flat on the metal surface (depicted in Fig. 5.2c). This sample shows inhomogeneity and at some positions, molecular clusters of more than 1 nm in height protruding from the molecular islands

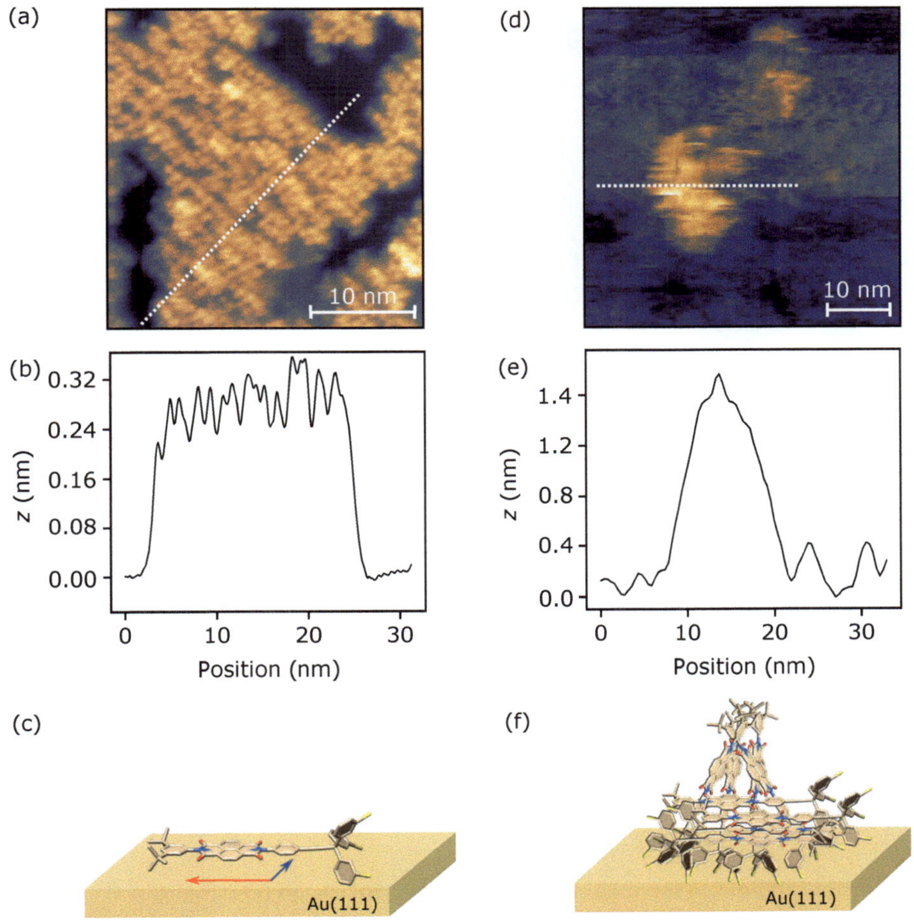

Figure 5.2: Topography of unordered clusters of Tpd-hNDI. (a) Typical topography of Tpd-hNDI on Au(111) with no long range order ($U = 1.8$ V, $I = 5$ pA). **(b)** Height profile taken along the white dotted line in **a** revealing apparent height of 0.3 nm. **(c)** Molecular model representing the possible orientation of the moelcule with long axis (red arrow) parallel to the substrate. **(d)** Two clusters of Tpd-hNDI, bright areas ($U = -2.3$ V, $I = 5$ pA). **(e)** Cross section along the white dotted line in **d** showing apparent height of more than 1.2 nm. **(f)** Cartoon representation of the molecular arrangement in a cluster.

are observed (see Fig. 5.2d and e). These protruding objects are > 8 nm wide, which suggests the formation of molecular clusters (depicted in Fig. 5.2f). The STM images of such clusters often show unstable behaviour, which is reflected by the white streaks and abrupt changes in the apparent height from one horizontal scan line to the next (see Fig. 5.2d).

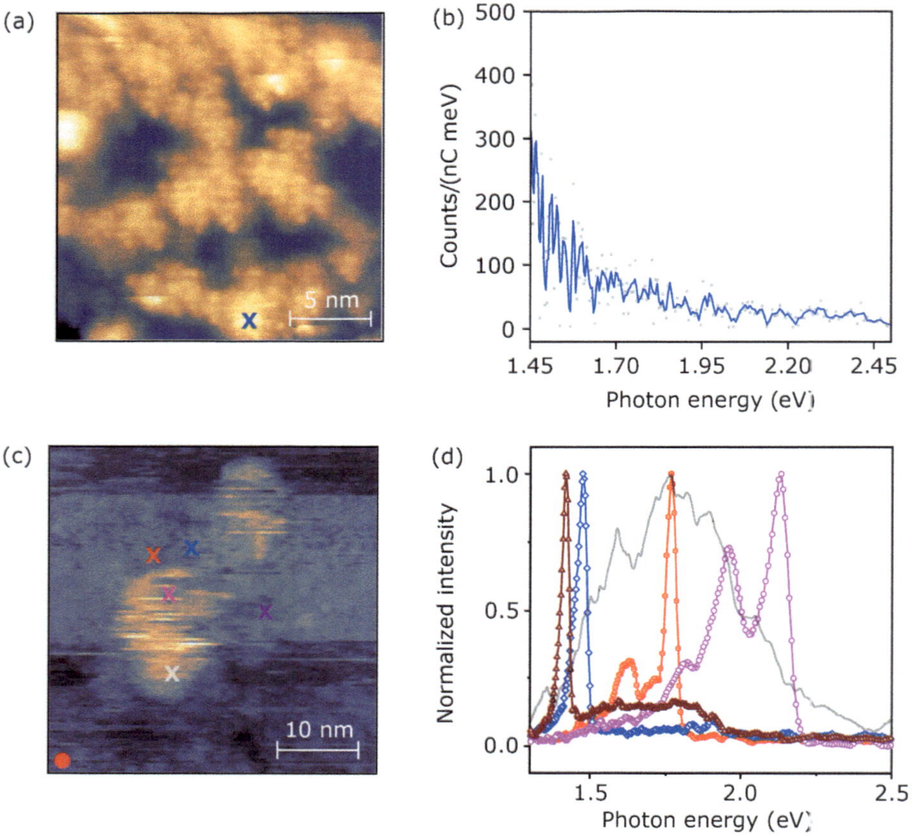

Figure 5.3: Optical emission spectra from unordered Tpd-hNDI islands (a) Topography of unordered islands of Tpd-hNDI ($U = 2.5\,\mathrm{V}$, $I = 3\,\mathrm{pA}$). **(b)** Optical emission spectrum measured at the blue cross in **a** shows no molecular luminescence ($U = 2.5\,\mathrm{V}$, $I = 120\,\mathrm{pA}$, $t = 1\,\mathrm{s}$). **(c)** Topography showing two clusters of Tpd-hNDI ($U = -2.3\,\mathrm{V}$, $I = 5\,\mathrm{pA}$). **(d)** Optical emission spectra recorded at positions marked (colored crosses) in **c**. Spectra are normalized with the maximum intensity in order to compare the spectra. Parameters for recording the emission spectra are $U = -2.5\,\mathrm{V}$, $I = 120\,\mathrm{pA}$, $t = 1\,\mathrm{s}$.

As per our expectation, no optical emission spectrum of molecular origin is observed from the islands shown in Fig. 5.2a. The molecules merely act as a dielectric layer (Fig. 5.3a) and partially or fully suppress the plasmonic light, which is the characteristic for noble metal nanogaps [11, 130], see Fig. 5.3b. However, on the clusters (Fig. 5.2d and 5.3c) and in their close proximity, many optical emission spectra show sharp peaks contrary to the typical plasmonic spectra. Figure 5.3c shows an area containing two molecular clusters on which 2500 emission spectra were recorded. At most positions, the photon spectra are similar to the grey spectrum shown in Fig. 5.3d recorded at the position indicated by the grey cross in Fig. 5.3c, with a broad peak (FWHM of 524 meV)

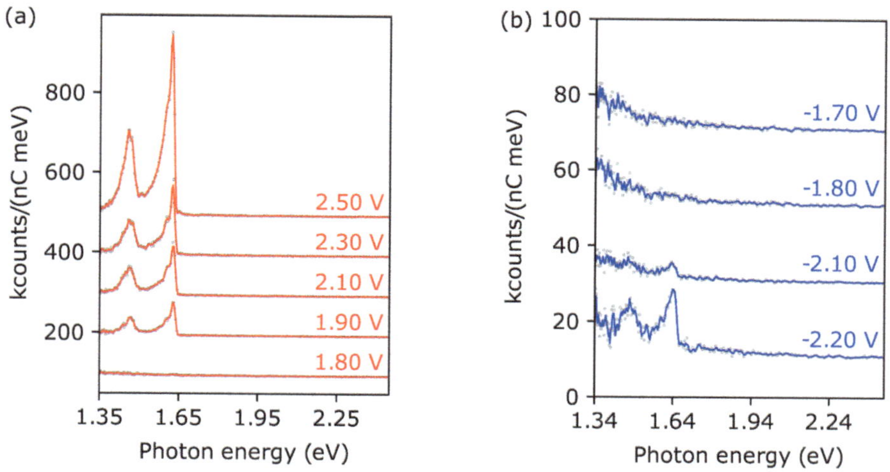

Figure 5.4: Energetics of emission from Tpd-hNDI. (a) and **(b)** Evolution of optical emission spectra with applied positive and negative sample bias voltages, respectively (I = 5 pA, t = 1 s).

centred around 1.75 eV which indicates towards plasmonic origin. Out of 2500 spectra, about 200 of them show much sharper peaks with FWHM between 30 meV and 100 meV. Some of these peaks are shown in Fig. 5.3d. It is interesting to see that the peak positions are distributed between 1.40 eV and 2.15 eV. The quantum yield, that is the number of photons emitted per tunnelling electron, ranges from 1×10^{-4} to 1×10^{-5} for an applied bias of −2.5 V. As mentioned before, the junction is not stable and for this reason it is important to normalize the counts by the average tunnelling current recorded during the measurement rather than using the set-point current. It is hard to say how these peaks are related to the exact measurement positions due to the sensitive nature of the spectra on tip positions. For example, the blue and red spectra are drastically different from each other and this difference cannot be inferred from the measurement positions (in topography). This can be explained by the absence of any long-range order which leads to a number of different electronic environments for each molecule. In the cluster, it is impossible to determine the electronic environment of each molecule, which drastically influences the emission energy. Nonetheless, at a fixed position, the tunnelling junction is stable enough to perform voltage dependence measurement which helps to understand the underlying excitation mechanism.

Figure 5.4a and b show the evolution of optical emission spectra with the applied positive and negative sample bias voltages, respectively. The measurement was done at a fixed position marked with a red dot in Fig. 5.3c. At the positive and negative polarities, the emission with peak energy 1.63 eV sets in above 1.8 V and below −2.1 V. This indicates an asymmetry of the junction under the reversal of the applied bias voltage. In the present measurement conditions, the high energy end of the plasmon peak is limited by

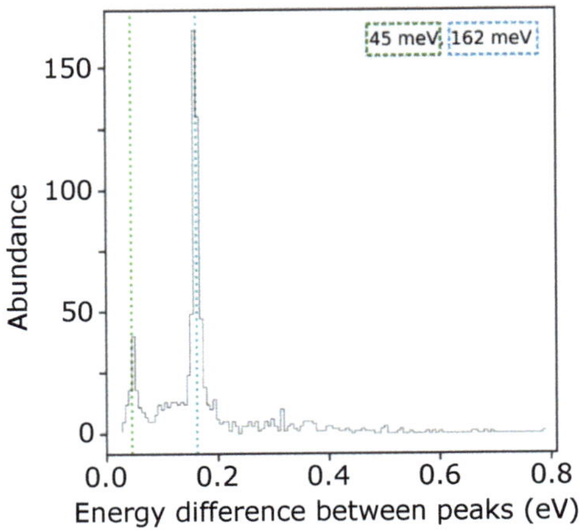

Figure 5.5: Histogram of energy differences between multiple peaks. The peaks are separated by 45 and 162 meV from the high energy peak of the spectra recorded on Tpd-hNDI clusters.

the energy of the tunnelling electrons, which means $eU = h\nu$. However, for the spectra in Fig. 5.4a and b, the threshold values for light emission observed are higher than the high energy edge of the emission spectra, which is a sign of molecular contribution in the light emission. Furthermore, two peaks are observed in the emission spectra in Fig. 5.4. It is interesting to see that both the peaks appear at the same applied bias voltage, which suggests that the low energy side peak belongs to the transition responsible for the peak at 1.63 eV. This side peak can be attributed to vibrational excitations [5, 22], which means $eU = h\nu + h\omega$, where ω is the energy of a vibrational mode. For most photon spectra recorded on the unordered clusters, the spacing between the two peaks is around $\sim$162 meV as shown in Fig. 5.5, where a histogram of energy difference between multiple peaks in photon spectra is plotted. It is important to mention here that on the unordered clusters of Tpd-hNDI, record high quantum yield of 1.3×10^{-3} photons/electron was observed [31], to best of knowledge [3, 20, 31] (see the spectrum in Fig. 5.4 recorded at 2.5 V). After correcting this value with the detector efficiency (see chapter 3), the internal quantum yield is 6×10^{-3} photons/electron.

Annealing the sample to higher temperatures promotes the diffusion of the molecules which in turn result in highly ordered molecular islands [31, 96, 125–129, 131]. For this purpose, Tpd-hNDI sample was further annealed to 180 °C. This leads to the formation of ordered structures as shown in Fig. 5.6a. Typical apparent heights of such islands are $\sim$350 pm (see Fig. 5.6b), which again suggests that the molecules indeed adsorb flat on the surface as depicted in Fig. 5.2c. A more closer image of the island in Fig. 5.6a is shown in Fig. 5.6c. The molecular structure is superimposed to scale to determine the possible molecular arrangement in the island. The proposed arrangement exhibits a

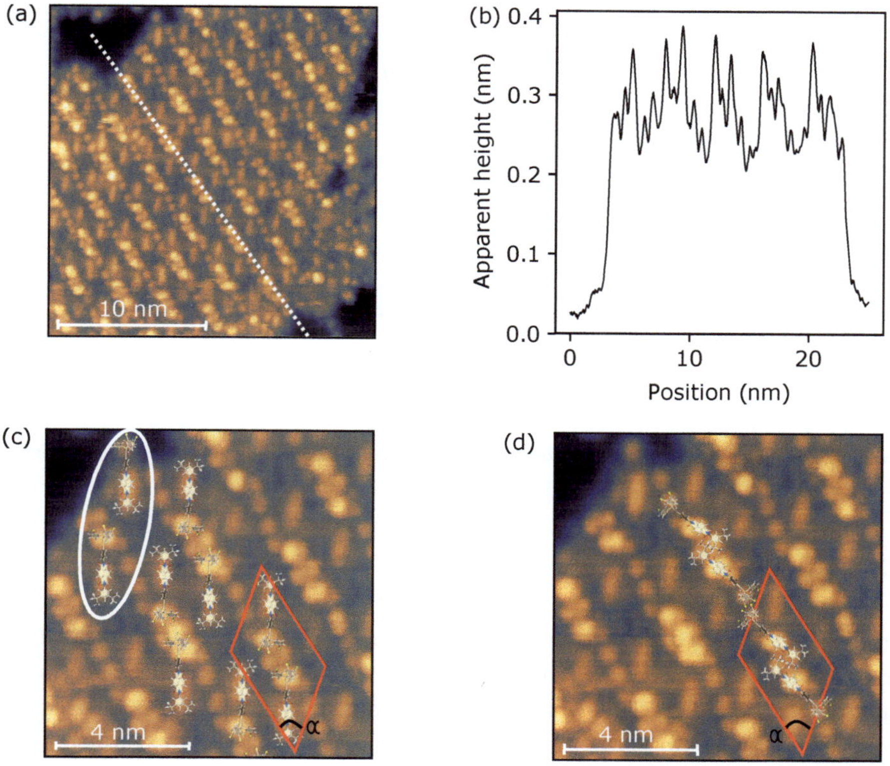

Figure 5.6: Ordered islands of Tpd-hNDI (a) Topography of an ordered island of Tpd-hNDI (U = 2.1 V, I = 9 pA). **(b)** Height profile along the white dotted line in **a.** **(c)** and **(d)** Islands with the unit cell indicated in red boxes. Molecular models of Tpd-hNDI superimposed to scale in the suggested two possible adsorption configurations [31].

monoclinic lattice structure with a unit cell of 2.7 nm $\times$ 4.04 nm = 8.71 nm^2, $\alpha = 53°$ (red lines). By repeating the unit cell, it is possible to fill the whole island (see the edges where a possible single molecule is present, white oval). Even though the proposed pair wise configuration holds nicely for the observed structure, it is still not unique and another configuration can also be proposed as shown in Fig. 5.6d [31]. Similar to the unordered islands (Fig. 5.2a), plasmonic light is suppressed on the ordered island in Fig. 5.6a. Figure 5.7 shows spectra recorded on the ordered island (in red) and on Au(111) (in black).

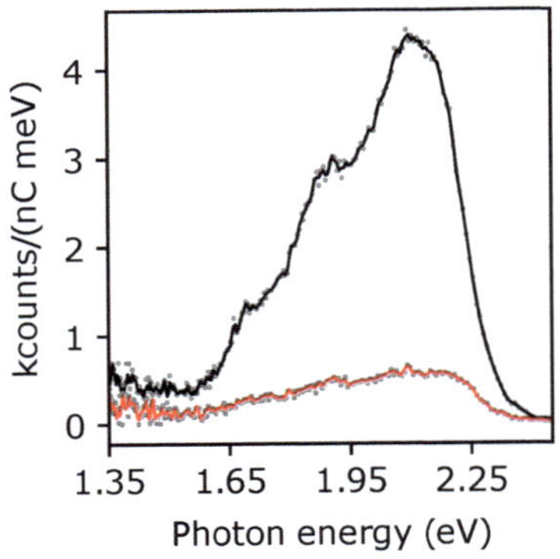

Figure 5.7: Optical emission spectrum recorded on Au(111) (in black) and on ordered Tpd-hNDI island (in red) ($U = 3$ V, $I = 140$ pA, $t = 1$ s) [31].

5.1.2 Tpd-sNDI Molecule

Sample with Tpd-sNDI was prepared in a similar way as the Tpd-hNDI. At the post annealing temperature of 100 °C, molecules form unordered islands and no clustering behaviour was observed. Due to the spray deposition, which may result in an inhomogeneous distribution of the molecules (see chapter 3), it is not possible to exclude the formation of clusters at this temperature. Figure 5.8a shows a typical topography of the unordered islands formed by Tpd-sNDIs. The apparent heights of these islands are 0.3 nm with some bright features showing heights more than 0.6 nm. These islands merely act as an dielectric layer in the junctions and suppress the intensity of expected plasmonic emissions (similar to Fig. 5.7).

After annealing the sample to ~180 °C, some molecular clusters were observed (similar to Tpd-hNDI) with the apparent height exceeding by 0.2 to 0.4 nm from its surrounding islands (see Fig. 5.8b). On these clusters, strong molecular fluorescence was observed. This can again be explained by the fact that these multilayer molecular islands have sufficient electronic decoupling from the metal surface such that the most protruding molecule is stable in the excited state long enough to emit light into the far field. Figure 5.8c shows two spectra recorded at the positions marked in Fig. 5.8b. In each of these spectra, two peaks are observed, both separated by 160 to 170 meV. The emission line width and spacing between the adjacent peaks are similar to the observations made on Tpd-hNDI (see Fig. 5.4a and b). Therefore, it is assumed in the case of Tpd-sNDI as well that the red shifted peak originates from a decay from the first electronically excited state to the vibrationally excited electronic ground state. Apparently, the local surrounding has a significant influence on both the electroluminescence efficiency

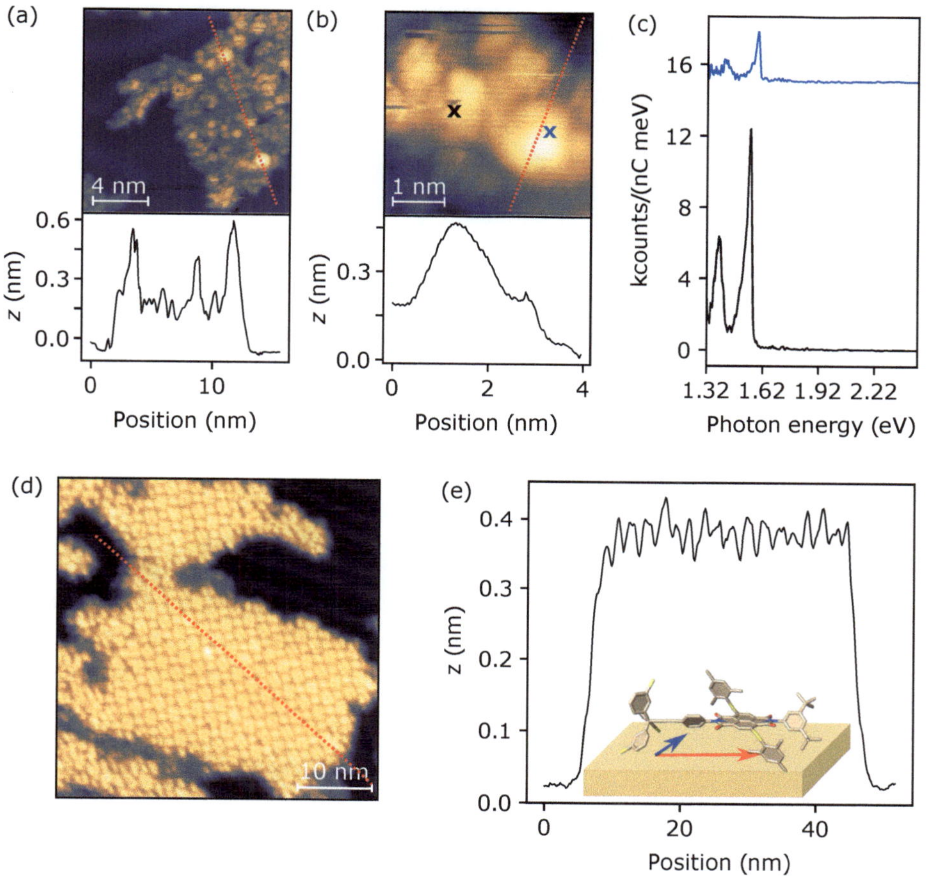

Figure 5.8: Topography and luminescence of as adsorbed Tpd-sNDI. (a) Unordered islands of Tpd-sNDI ($U = -1.4$ V, $I = 2$ pA). Bottom panel shows the height profile along the red dotted line in the top panel. **(b)** Cluster of Tpd-sNDI ($U = 2.4$ V, $I = 5$ pA) with height profile taken along the red dotted line (bottom panel). **(c)** Optical emission spectra recorded on two positions marked with corresponding colored crosses in **b**. Parameters for recording the spectra are $U = 2.5$ V, $I = 25$ pA, $t = 2$ s [31]. **(d)** Ordered island of Tpd-sNDI is fromed once the sample is annealled at higher temperatures ($\sim$180 K) ($U = 1.8$ V, $I = 3$ pA). Height profile along the red dotted line shows apparent height of 0.3 to 0.4 nm. Schematic of molecules lying flat on Au(111) surface (not to scale) with long axis (red arrow) parallel to the substrate [31].

and the energy of the emitted light. Emission efficiency is in the order of $\sim 10^{-4}$ photons/electron. In agreement with the recent study on PDI tripodal molecules [17], the molecular configuration seems to be easily influenced by the nearby tip which hampers the reproducibility of STML experiments.

At elevated annealing temperature ($\sim$180 °C), similar to Tpd-hNDI, Tpd-sNDI also forms well-ordered islands as shown in Fig. 5.8d. Apparent height of 0.3 nm suggests that these

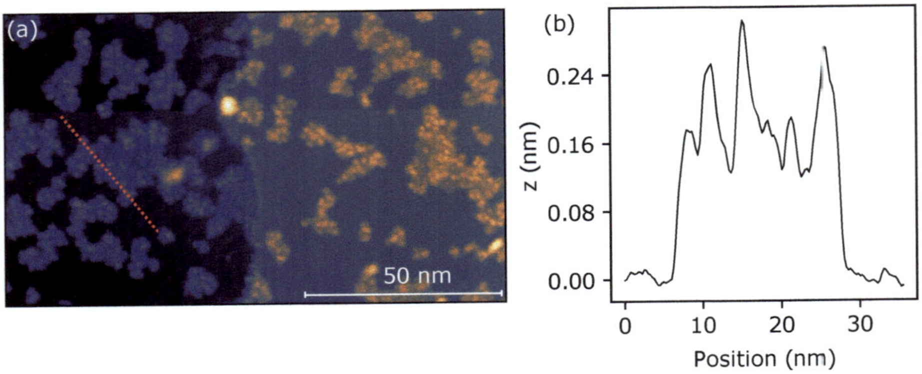

Figure 5.9: Topography of as adsorbed Tpd-nNDI. (a) Unordered islands of Tpd-nNDI on Au(111) sufrace ($U = -1.9\,\text{V}$, $I = 2\,\text{pA}$). **(b)** Height profile shows the apparent height of 0.3 nm.

islands are also composed of flat lying molecules (see Fig. 5.8e). On these islands, the emission spectra do not show any molecular fluorescence but only suppressed plasmon intensity is observed, these molecules, however, show the interesting phenomenon of switching under the influence of STM tip. This switching also changes the emission spectra of these molecules. Switching and respective light emission properties will be discussed in the section 5.2 in all details.

5.1.3 Tpd-nNDI Molecule

At the post annealing temperature of 100 °C, Tpd-nNDI molecules also lie flat on the Au(111) surface with the molecular long axis parallel to the substrate (see Fig. 5.2c). Molecules form islands on the surface with no long range order as shown in Fig. 5.9. As expected, molecular fluorescence was quenched and indeed no molecular features in the photon spectra were observed. Also, for these molecules, clusters consisting of molecules in a more upright configuration were not observed.

5.2 Switching Tpd-sNDI on a Metal Substrate to a Luminescent State

It was discussed in the section 5.1.2 that once the Tpd-sNDI sample was annealed at higher temperatures (~ 180 °C), ordered islands were observed. Interestingly, even though the molecules did not show any fluorescence in their pristine state after deposition, it was still possible to selectively switch individual molecules within the ordered

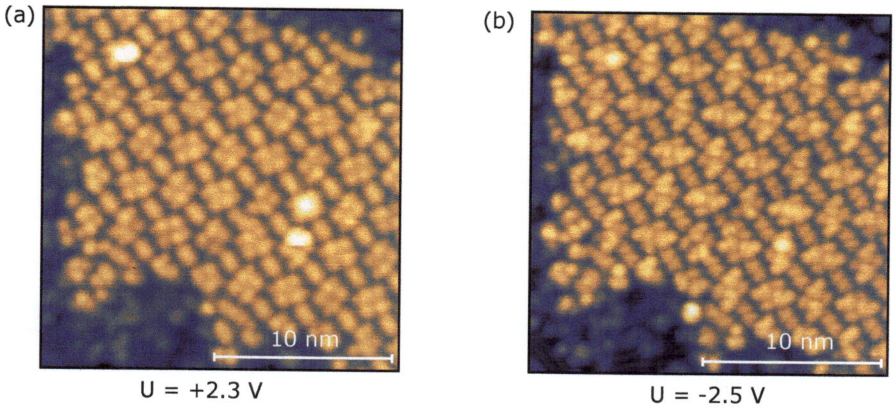

Figure 5.10: (a and b) Ordered island of Tpd-sNDI scanned at positive (= 2.3 V) and negative (= −2.5 V) sample voltages. Tunnelling current for recording the images I = 12 pA.

islands into a light emitting state. This again shows that the STM is a capable technique to switch and modify individual molecules [125–129, 132, 133].These results provide the first proof of principle of well-defined electrically induced light emission from self-decoupled molecules directly adsorbed on a metal surface.

5.2.1 Adsorption Configuration

Figure 5.10a and b show two STM images of the same sample area recorded at opposite bias voltage polarities. Both scans show slightly different topographies. This suggests that different states of the molecules are involved in the tunnelling process at both polarities. To determine the adsorption configuration of the molecules in such ordered islands, an enlarged view is shown in Fig. 5.11b where Fig. 5.11a shows the molecular structure. Here, similar to Tpd-hNDI, a pair wise adsorption configuration of the Tpd-sNDI is proposed. Molecules of adjacent chains are mirrored (see two adjacent pairs in Fig. 5.11b). This suggests a rather tightly packed configuration with the unit cell of area 3.3 nm × 5.0 nm ($\alpha = 47.6^0$) = 12.2 nm^2 which contains 4 molecules (area = 3.05 nm^2/molecule). To further support this hypothesis, Fig. 5.12a shows a pair of Tpd-sNDI with red and green contours superimposed on them to distinguish between the two adjacent molecules in the pair. If the proposed hypothesis is correct, then by repeating the unit cell shown in Fig.5.11b should cover the whole island. To test this, an Tpd-sNDI island was considered (in Fig. 5.12b) in which the edges of the island are clearly visible. By superimposing the contour shown in 5.12a over the island, the edges of the molecular island (white arrows) show one quarter of the unit cell as the smallest unit,

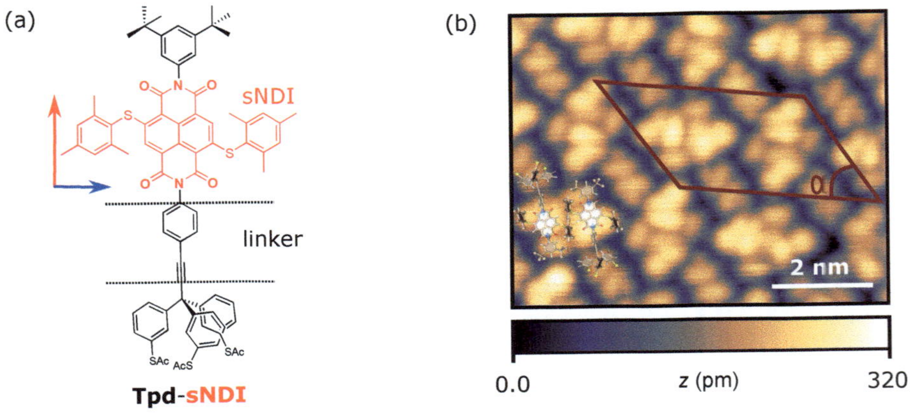

Figure 5.11: Topography of Tpd-sNDI. (a) Chemical structure of Tpd-sNDI as synthesized, comprising a chromophore (red) and a tripodal anchor (black). Long and short axis of the molecule is represented by red and blue arrow, respectively. **(b)** STM image of Tpd-sNDI molecules in the pristine state ($U = -2.5$ V, $I = 12$ pA) with molecular structure superimposed to scale. The monoclinic unit cell is represented in brown parallelogram containing four molecules.

which confirms the hypothesis of pairwise adsorption with adjacent molecular chains mirrored (see chains 1 and 2 in Fig. 5.12b).

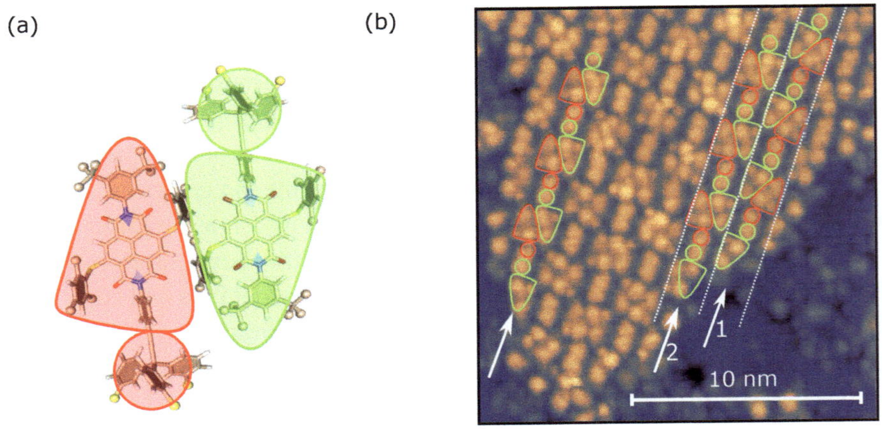

Figure 5.12: Adsorption configuration of Tpd-sNDI molecules on Au(111). (a) A pair of Tpd-sNDI with red and green contours superimposed on them. **(b)** Contours shown in **a** are suprimposed on the STM image of a Tpd-sNDI island. White arrow represent the molecules on the edges of the island.

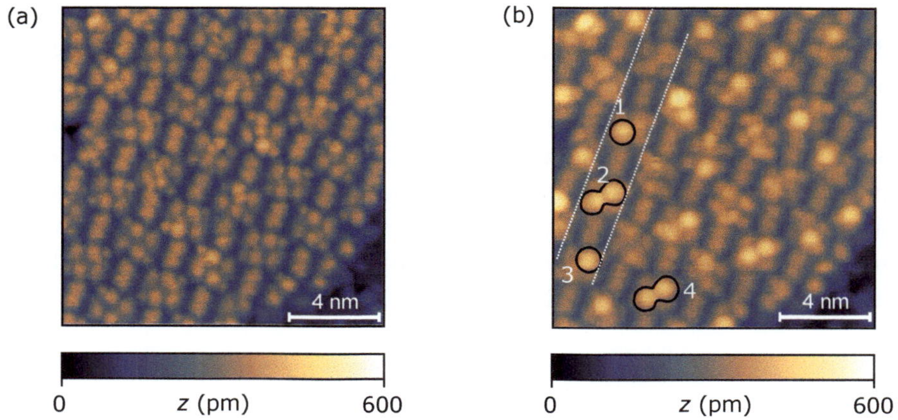

Figure 5.13: Switching of Tpd-sNDI. (a) STM image of flat lying Tpd-sNDI molecules on Au(111) surface (U = 1.8 V, I = 2 pA). **(b)** By scanning the area in **a** at high positive sample bias voltage (U = 2.5 V), molecules can be switched into a state with increased apparent height. Tunnelling parameters for recording the image are $U = -2.4$ V, I = 2 pA.

5.2.2 Switching

By applying high positive sample bias voltages ($U > 2.5$ V), individual molecules can be switched such that they display an increased apparent height in the STM images. Figure 5.13a shows an ordered island of Tpd-sNDI. After scanning this area at high bias (U = 2.5 V), several of the molecules appear higher in the topography, for example, molecules marked with **1, 2,3** and **4** in Fig. 5.13b. It is interesting to note that in the pair of molecules (as in Fig. 5.11), it is possible to switch one (as **1** and **3**) or both molecules (as **2** and **4**). This switching process is highly reproducible and reversible, which allows to write the letters "INT" on an ordered monolayer of Tpd-sNDI (see Fig. 5.14 top panel). As expected, flat lying molecules do not show any molecular fluorescence [11, 134].

Besides the change in apparent height, a fraction of the switched molecules also show a drastically different optical emission spectrum (see Figure 5.15) corresponding to a molecular pixel switched into the fluorescent state. Even though this is a proof of the concept of an ultimate resolution of a molecular display, not all the switched molecules can be driven into the light-emitting state. This prohibits writing "INT" with light emitting molecules. In contrast to prior attempts reported in the literature to self-decoupled molecules [14, 15, 17, 31], well-reproducible optical emission spectra with sharp emission lines were observed, including vibrational side bands. Moreover, once the molecules are switched into the light-emitting state, they are very stable and robust. Figure 5.15a shows an island with several molecules switched (bright features). Among them, the molecules marked with a blue cross show emission spectrum with sharp peaks and rest of the molecules switched/unswitched only show suppressed plasmon intensity compared to Au(111) (see spectra in Fig. 5.15b corresponding to the crosses in Fig. 5.15a). The optical emission spectrum (blue) in Fig. 5.15b shows a bright main peak (X^+) at ~ 1.88 eV, followed by a series of lower intensity peaks at lower energies (see Figs. 5.15b and 5.16a). The observed quantum efficiency of about 10^{-5} - 10^{-4} photons/electron is similar to light emission from individual molecules deposited on insulating layers [3]. Surprisingly, the energy of the X^+ line does not match our photoluminescence measurements of Tpd-sNDI in solution [31], which suggests that it is not arising from an excited neutral state of Tpd-sNDI.

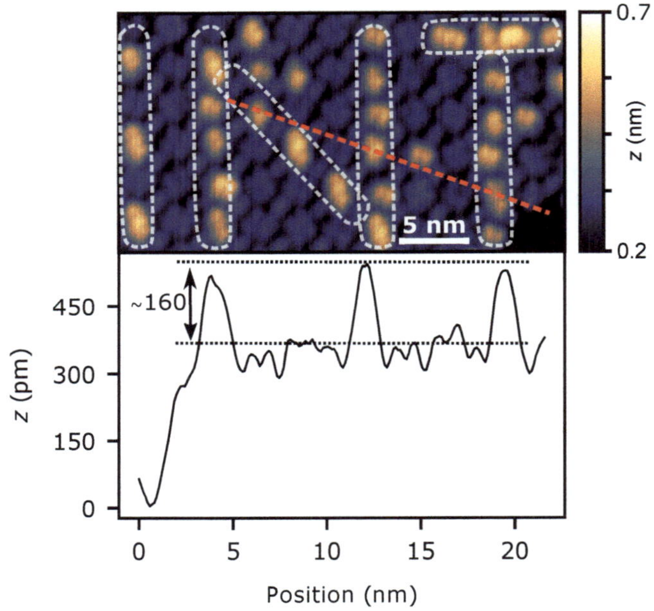

Figure 5.14: Top, letters "INT" written on an ordered island of Tpd-sNDI by controlled manipulation with the STM tip (U = 2.5 V, I = 3 pA). **Bottom**, measured apparent height along the red dotted line marked in top panel. Switched molecules appear ~160 pm higher than the flat lying molecules.

To understand the origin of the X^+ line, quantum chemical calculations of neutral and charged Tpd-sNDI molecules were performed. Theoretical calculations were performed by our collaborators from the group of Prof. Carsten Rockstuhl at *Institute of Theoretical Solid State Physics*, KIT. The calculations for neutral forms of Tpd-sNDI monomers and dimers show emission lines at much higher photon energies than the experiment. For the neutral Tpd-sNDI molecule, no emission band below 2.0 eV could be detected that is able to cause the observed light emission at ~1.9 eV. Instead, for the monomer a single excited state located at 2.72 eV is found by TD-CAM-B3LYP [112, 113, 135, 136]. Dimerization alone also only induces a minor shift compared to a single Tpd-sNDI

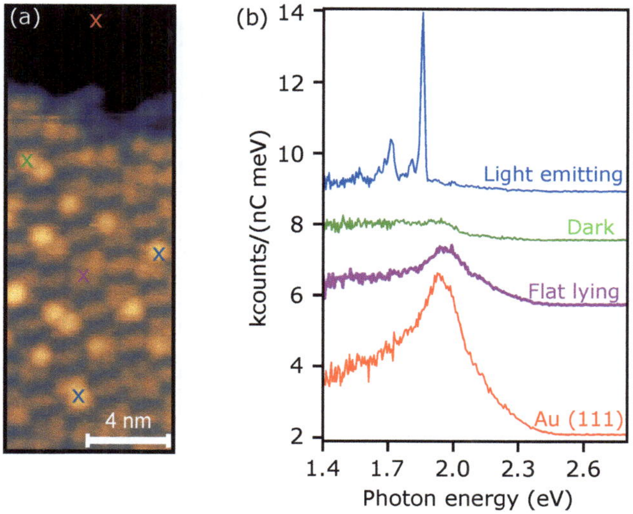

Figure 5.15: **(a)** STM image of an ordered island with some molecules switched. Image is recorded with tunnelling parameters of $U = -2.3\,\text{V}$, $I = 2\,\text{pA}$. **(b)** Spectra recorded by placing STM tip on different molecular motifs and on Au(111). Positions are marked with colored cross in **a** for the respective spectra. The spectrum on Au(111) shows typical broad plasmon peak centered around 2 eV (in red). On the flat lying molecule, the plasmon spectrum is suppressed (in purple). On the switched molecule which do not emit light (dark), the plasmon spectrum is suppressed even further (in green). The switched molecules, which do emit light, sharp molecular peak appears at 1.88 eV (in blue). Measurement conditions for all the spectra are $U = -2.50\,\text{V}$, $I = 30\,\text{pA}$, $t = 3\,\text{s}$.

molecule. Two excited states with nearly the same energy are found at 2.69 and 2.73 eV. This behaviour can be expected from mainly non-covalently bound dimers. Neither a single Tpd-sNDI molecule nor the neutral dimer alone can account for the observed occurrence of emission peaks at 1.90 eV. Therefore, charging processes have to be taken into account to explain the observed light emission.

5.2.3 Energetics

Once TD-CAM-B3LYP calculations were done for the charged dimer, gas phase spectrum matches the energy of the observed X^+ line (see Fig. 5.16a and b). This excited state can be related to an intra-molecular charge-transfer excitation. The hypothesis of the formation of charged Tpd-sNDI dimers is also in good agreement with the proposed pairwise adsorption configuration (as shown in Figs. 5.11 and 5.12) [31]. The charging of the molecule is expected to depend on the polarity of the applied bias [23, 29]. Indeed, the X^+ line only appears at negative sample bias (see Figure 5.16c). The critical sample

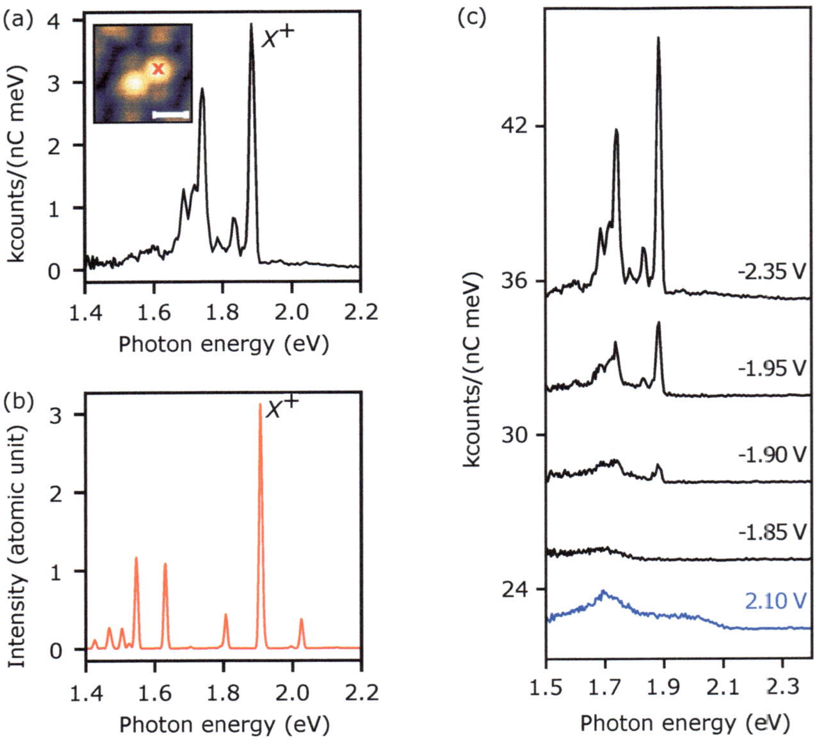

Figure 5.16: Excitation of charged Tpd-sNDI dimer. (a) Optical emission spectrum of the Tpd-sNDI dimer with the STM tip positioned at the red cross in the inset topography ($U = -2.35$ V, $I = 6$ pA, $t = 60$ s). Scale bar in the inset image is 1 nm. **(b)** Simulated TD-CAM-B3LYP photoabsorption spectrum of charged Tpd-sNDI dimer with all electronic transitions in the indicated energy range. **(c)** Optical emission spectra at positive (blue) and negative (black) sample biases ($I = 6$ pA, $t = 60$ s).

bias for the onset of the X^+ line matches its energy very closely, which indicates the excitation of the molecule by inelastic energy transfer (IET) via plasmons [6, 137] (see chapter 2). All peaks at lower energies are set in at the same sample bias (see Figure 5.16c), and can be assigned to the vibrational satellites of the X^+ line [5, 22, 63].

5.2.4 Spatial Variation

The X^+ line is observed only when the tip is placed on top of chromophore part of the molecule. This is more clearly visible in the photon map measured on a dimer shown in Fig. 5.17. Figure 5.17a shows a switched Tpd-sNDI dimer with the suggested molecular structure superimposed to scale. Figure 5.17b shows the corresponding spatial

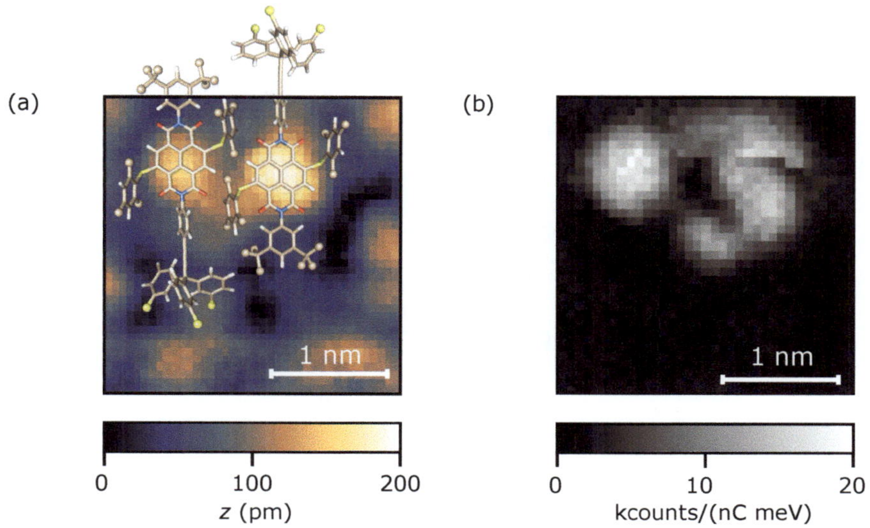

Figure 5.17: Spatial variation of X^+ emission line. (a) Topography of switched Tpd-sNDI dimer with molecular structure superimposed to scale. **(b)** Simultaneously recorded intensity of the X^+ line (grid of 40×40 points, integrated photon count in the energy range of 1.86 eV to 1.92 eV), $U = -2.3$ V, $I = 5.1$ pA, $t = 1$ s.

distribution of the intensity of the X^+ line recorded on a grid of 40×40 points. It is clear that the light is emitted into the far-field as soon as the tip is placed directly on top of one of the two bright features present in the topography. This reflects that the emission of X^+ is related to the electronic states visible in the topography of the dimers while scanning at negative sample bias voltage. A more detailed discussion of the electronic structure of the molecule is done in the following section.

5.2.5 Electronic Structure

For further understanding of the light emission from different molecular configurations (flat molecules, switched molecules with and without strong molecular luminescence signal), a systematic study of electronic structure is done from two perspectives; First, by STM measurements and second with the help of DFT calculations.

I. STM Measurements:

Normalized differential conductance $(dI/dU)/(I/U)$, which is proportional to the local density of states (LDOS), was recorded upon switching Tpd-sNDI into different states (see Fig. 5.18). The STM tip was unchanged during these measurements (see black spectrum in Fig. 5.18), ensuring that the contrast in the spectra measured on the molecules

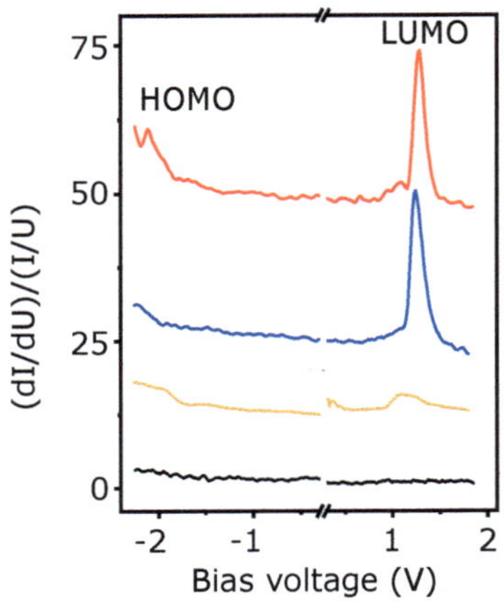

Figure 5.18: STS on different states of Tpd-sNDI molecules. Normalized dI/dU spectra measured on Au(111) (black line), on a flat-lying pristine molecule (yellow line), on an switched molecule which shows no molecular luminescence (blue line) and on a switched fluorescent molecule (red line).

is due to the involvement of different electronic states. On a flat lying molecule, only a broad peak at $\sim$1.11 V is observed (in yellow), whereas on a switched molecule this peak appears much sharper (in blue and red). On a switched molecule, which does not emit light, peak in positive bias is at $\sim$1.24 V (in blue). For this molecule, no peak is observed in the negative bias. On the switched fluorescent molecule, peak in positive bias is centered around $\sim$1.29 V (in red). This suggests that the observed change in the topography of the molecules once they are switched (see Figs. 5.15a and 5.14) goes along with a certain electronic decoupling of the corresponding orbital of the molecule, which is tentatively identified as the lowest unoccupied molecular orbital (LUMO). In the negative bias, the fluorescent molecule shows another peak at ~ -2.11 V, which has an onset at ~ -1.70 V. This frontier orbital can be assigned to the highest occupied molecular orbital (HOMO), however, calculations (see the next subsection) show that the density of states below the HOMO is high, this actually leads to many possible excitations - which have in common that the final state is LUMO, but the hole orbital (HOMO) has many possibilities. The distribution of these orbitals can also be viewed in the topography of the molecules, however, it is not possible to distinguish the switched molecules which emit light from those which do not (see Figs. 5.13b and 5.15a). Figure 5.19a shows a part of Tpd-sNDI island with all molecules in the flat state. A small area of this island (masked rectangle) is shown in Fig. 5.19b and c, where three molecules are switched (marked as 1,2 and 3) by applying high positive bias (U = 2.5 V). In topography measurements at positive and negative sample biases, the molecules show different motifs (red and black contours in Figure 5.19b and c) that can be associated with the

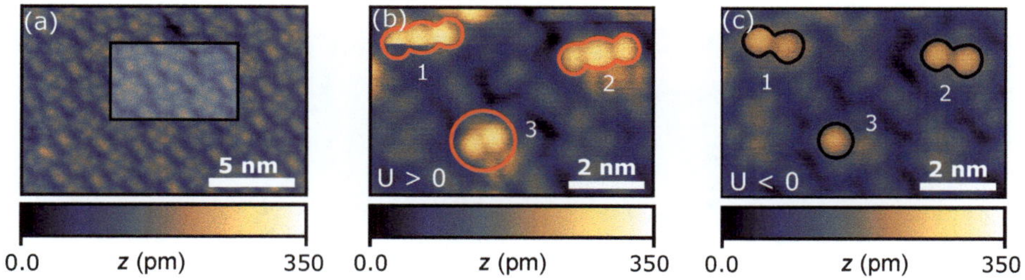

Figure 5.19: (a) STM image of an ordered island of Tpd-sNDI recorded at U = 1.8 V, I = 2.4 pA. **(b)** and **(c)** Zoomed image of the rectangular area marked in **a**, recorded at positive bias (U = 2.5 V, I = 2.1 pA) and at negative bias (U = −2.4 V, I = 2.1 pA) voltages after switching. Molecular motifs of the switched molecules (labelled 1, 2, and 3) are marked with red and black contours.

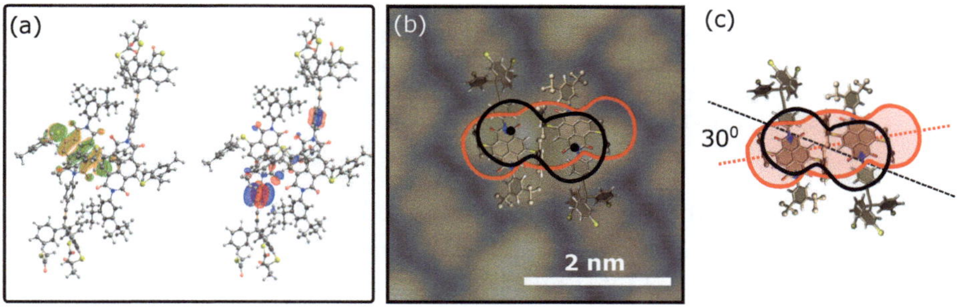

Figure 5.20: (a) Particle (left) and hole (right) natural transition orbitals (NTO) involved in the main optical transitions of the charged dimer. **(b) and (c)** Molecular models superimposed to the topography of a pair of Tpd-sNDI molecules together with white and black contours of the switched molecule taken from Figs. 5.19b and c.

LUMO and HOMO, respectively. It is worth noticing that the discussed molecules (1,2 and 3) are not emitting light into the far-field.

II. Theoretical Analysis:

It was discussed in the section 5.2.3 that only the excitation of the charged dimer gives the observed light emission (Fig. 5.16c). For further understanding of the distribution of the orbitals (HOMO and LUMO) discussed in the above section, the spatial extent of the calculated NTOs of the X^+ line for the excitation of the charged Tpd-sNDI dimer was extracted. It is important to mention that this transformation (NTO) is particularly useful when working with the excited states of molecules with multiple chromophoric sites (as for Tpd-sNDI dimer), which provides a more intuitive picture of the orbitals, (mixed or not), involved in any hole-particle excitation [138].

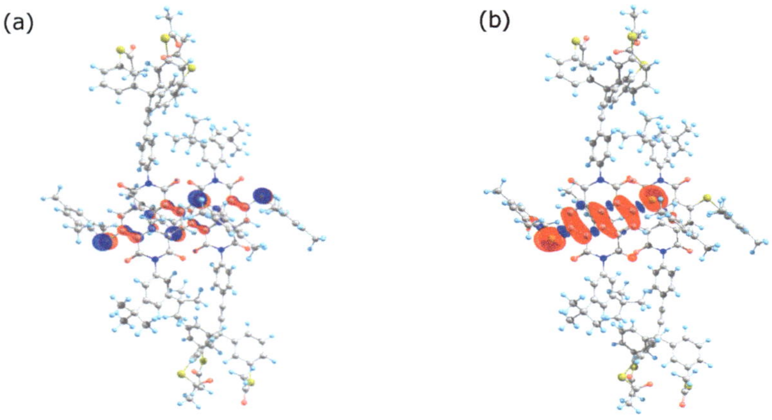

Figure 5.21: (a) Highest occupied molecular orbital (HOMO) of the neutral dimeric complex, delocalized over both NDI centers in the dimer. **(b)** Spin density of the cationic dimer assembly, which is spatially nearly identical to the LUMO, localized at a single NDI center in the dimer. Plots obtained at the CAM-B3LYP/def2-TZVP (def2-SVP for H) level of theory.

In good agreement with the molecular motif assigned to the LUMO in Fig. 5.19b, calculated particle NTO (Fig. 5.20a, left) is centered on the sNDI chromophore and extends further out to the arylsulfanyl substituent. For the assigned HOMO orbital in Fig. 5.19c, the hole NTO is located mainly at the edge of the chromophore and extends to the phenyl ring between the chromophore and the tetraphenylmethane (Fig. 5.20a, right). Figure 5.20b and c shows the distribution of the assigned HOMO and LUMO (from the topography) with respect to each other. This agreement with the theory justifies the observation, *i.e.*, removing one electron is only possible when tunneling into the HOMO [29] as clearly visible in the photon map (Fig. 5.17b).

As mentioned earlier , switching of the dimers alone would not lead to the observed fluorescence of the molecules. For the light emission, charging is necessary (DFT calculations, see Fig. 5.16b). For further understanding of the charging process, HOMO distribution of the dimer was calculated (shown in the Fig. 5.21a), which reveals that the HOMO is delocalized over both the NDI molecules. This suggests, charging of the dimer can possibly occur at both the NDI centers. Whereas, the spin density of the charged dimer (see Fig. 5.21b) is only located at one of the two NDI centers. Therefore, it is more likely to charge the dimer than the monomer given that charging can occur at both the molecules. Apart from that, due to the localization of the spin density on one NDI center, the unpaired electron can localize on the more stable center, which is the electronically more decoupled switched molecule apparent in the topography (Fig. 5.13). Both of these facts further support the model of light emission from the charged dimer of Tpd-sNDI. These calculations also suggests that even though the formation of

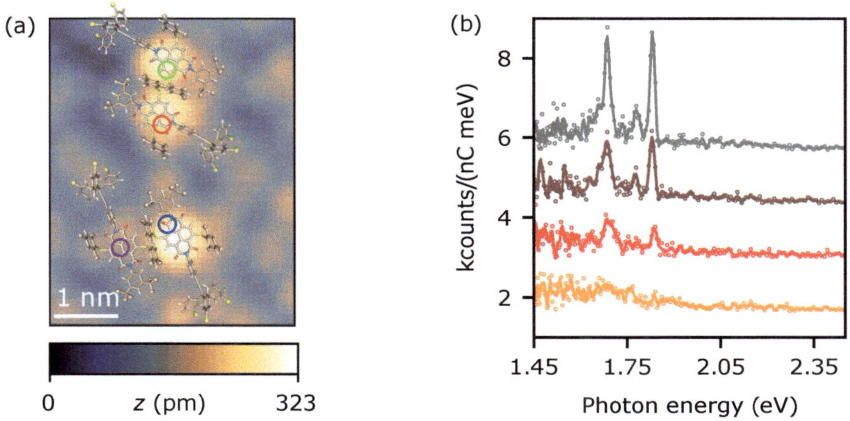

Figure 5.22: Excitation of pair of charged Tpd-sNDI dimers. (a) Topography of two adjacent dimers among which 3 molecules are switched (three bright features). Tunnelling parameters for recording the image are $U = -2.4\,\mathrm{V}$, $I = 3\,\mathrm{pA}$. **(b)** Optical emission spectra recorded at the positions marked with corresponding colored circles in **a** ($U = -2.4\,\mathrm{V}$, $I = 8\,\mathrm{pA}$, $t = 3\,\mathrm{s}$).

the dimer is necessary for the light emission, it should still be possible to only decouple one of the molecules in a dimer (where the spin density/unpaired electron can localize) and expect light emission. In fact, we do observe such behaviour as shown in Fig. 5.22, where two dimers are located adjacent to each other, among them three molecules are switched and all three of them are emitting light into the far-field. Moreover, the NTO analysis of the main emission line located at 1.88 eV (shown in Figure 5.16b) reveals that this excitation is an inter-molecular charge-transfer excitation. While the quantum chemical calculations allow to identify the orbitals that leads to the observed molecular light emission, the precise structural change that decouples the molecule is far more complex.

5.2.6 Underlying Mechanism

Figure 5.16c clearly shows that the onset energy matches the peak energy of X^+ emission line ($eU = h\nu$), which suggests strongly to the fact that the molecule is excited by the decaying gap plasmons (IET mechanism, also see chapter 2). Moreover, charging of the molecule is only possible by removing an electron from the HOMO orbitals, which means the HOMO is in resonance with the Fermi energy of the sample to charge the molecule. The onset of the HOMO peak in the dI/dU experiment at a sample bias of $-1.70\,\mathrm{V}$ and at this voltage, charging by removing an electron starts to be possible. However,

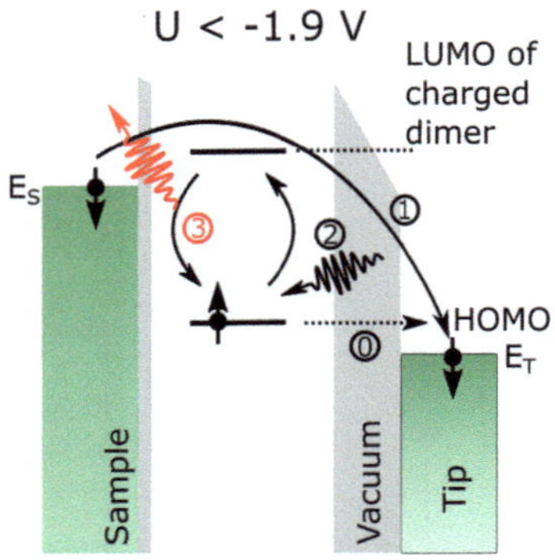

Figure 5.23: Excitation mechanism. (a), Mechanism for X^+ emission line. At sufficiently high bias, the electrochemical potential (E_T) of the tip is lower than the HOMO of the molecule and an electron is transferred from the HOMO to the tip hence charging the molecule (process 0). The LUMO of the charged molecule is now $\sim$ 1.90 eV above the HOMO. A plasmon is excited by an inelastically tunnelling electron (process 1) and in turn excites the molecule (black wavy arrow, process 2). The Excited molecule decays by emitting a photon into the far-field (red wavy arrow, process 3).

light emission only sets in at -1.9 V (see Figure 5.16c). Based on these observations, a phenomenological model can be proposed as shown in Fig. 5.23.

5.3 Summary and Conclusion

In this chapter, light emission from three different chromophores, namely, Tpd-hNDI, Tpd-nNDI, and Tpd-sNDI were studied. These chromophores were anchored to a tripodal platform to decouple them from the metal surface underneath. For the deposition of these molecules, the spray deposition technique was used. Post annealing of the sprayed sample at different temperatures resulted in molecules forming different assemblies. Contrary to our expectations, molecules lie flat on the surface and form molecular clusters or ordered/unordered islands [31].

For Tpd-hNDI, annealing at 100 °C lead to the formation of clusters and unordered islands. From the molecular clusters, light emission spectra of molecular origin were observed. However, due to the ill-defined electronic environment, molecular peaks vary in large energy range depending on the measurement positions. No molecular emission was observed from the unordered islands. Further annealing the sample to 180 °C, well ordered islands were observed and molecular fluorescence was quenched [31].

Tpd-nNDI molecules only formed unordered islands once annealed at 100 °C. No clustering of molecules were observed. Light emission spectra only showed suppressed plasmons [31].

The most interesting behaviour was observed for the Tpd-sNDI. Samples annealed at 100 °C show formation of molecular clusters. Optical emission spectra recorded on these clusters showed very similar behaviour as Tpd-hNDI. However, once the sample was annealed to 180 °C, molecules formed ordered islands [31]. Molecules in the island can be switched into different electronic states with the help of the STM tip. Contrary to the expectations, some of the switched molecules showed drastically different emission spectra with sharp peaks. Application of high positive sample bias decouples the LUMO of individual molecules from the electron bath of the underlying metal. This switching is observed as a surprisingly small change in the apparent height of the molecules. In some of the molecules, both HOMO and LUMO are decoupled which goes along with an increased lifetime of the excited state of the molecule such that photon emission with typical quantum yields becomes possible. Quantum chemical simulations of the gas phase spectrum revealed that the observed photon emission originates from the positively charged dimer of the Tpd-sNDI molecule.

In conclusion, measurements on these molecules suggest that even a small decoupling can lift the fluorescence quenching caused by the metal substrate. These observations provide guidance towards future approaches and modifications that are needed to achieve fully self-decoupled light emitting molecules. In essence, the anchored group can be extended, so that the molecule can overcome the van der Waal forces such that the molecules do not lie flat on the metal surface. This approach has been tested and the obtained results are promising and will be discussed in the next chapter.

6 Light Emission from Single Self-decoupled Molecules

In the previous chapter, light emission from Tpd-hNDI, Tpd-nNDI, and Tpd-sNDI has been studied. It was clear from the STM and STML measurements that the molecules do not stand in the upright configuration. Instead, they prefer to lie flat on the metal surface, hence quenching the molecular fluorescence. Anchoring the chromophores has been a challenging task since molecular anchors not only have to provide sufficient electronic decoupling of the light emitting chromophores but also should bond strongly to the metal electrode. For this reason, it is necessary to explore different anchoring groups to achieve a self-decoupled and rigid light-emitting single molecule.

In this chapter, light emission from two chromophores studied in the previous chapter, -sNDI and -nNDI, anchored to an extended tripodal platform (Tpdx) will be discussed. The two molecules are named as Tpdx-sNDI and Tpdx-nNDI. Figure 6.1a and b show their chemical structures, where the chromophores are highlighted in red and blue, respectively. Chromophores are linked to the tripodal platform via a phenylacetylene linker (Fig. 6.1a and b). These two molecules were synthesized by our collaborators from the group of Prof. Dr. Marcel Mayor at the *Institute of Nanotechnology*, KIT. Detailed structure of the tripodal platform has been discussed in chapter-2 [139]. Figure 6.1c shows the schematic of the expected molecular adsorption geometry (not to scale), with three thiol anchoring groups in the meta-positions bonded to the Au(111) surface and the possible upright orientation of the protruding chromophore (-sNDI).

This is among the few studies where light emission from such self-decoupled molecules has been studied in a well-controlled and reproducible manner. This chapter will start with a detailed discussion of STM and STML experiments on Tpdx-sNDI, followed by the experiments on Tpdx-nNDI. Finally, the correlation between their emissions will be discussed.

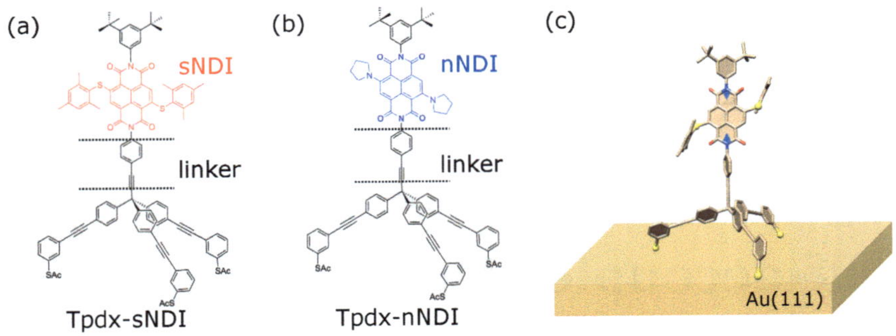

Figure 6.1: (a and b) Chemical structures of Tpdx-sNDI and Tpdx-nNDI, constituting the tripodal platform (in black) and the light emitting chromophores (in red and blue respectively). **(c)** Schematic showing upright configuration of Tpdx-sNDI on Au(111).

6.1 Experiments on Tpdx-sNDI

6.1.1 Sample Preparation and Topography

Deposition of Tpdx-sNDI on a clean Au(111) surface was done by the spray deposition technique (see chapter 3) [96]. For the deposition, 1 μL solution of Tpdx-sNDI in CH_2Cl_2 was sprayed on Au(111) followed by post-annealing of the samples at 100 °C. Figure 6.2a shows a typical STM image of the post-annealed sample. The topography shows several high objects with apparent height and width (FWHM) of $\sim$ 730 pm and $\sim$ 1.84 nm as shown in Fig. 6.2b (Gaussian peak profile fit, see cross section along the red dotted line in Fig. 6.2a). The estimated dimensions of the Tpdx-sNDIs in the minimum energy configuration (MM2 minimization in ChemOffice) are $\sim$ 2.56 nm and $\sim$ 1.76 nm in height and width, respectively. In STM, the real height of the molecules is only accessible indirectly [123], however, consistency between the measured width and calculated width hints that these high objects could be Tpdx-sNDI. Moreover, the isolated nature of the adsorbed objects suggests that Tpdx platform is anchored to the substrate. There are also residues from the sprayed solution, as well as the contamination from the gas phase which are also visible in the STM image with much lower apparent height (< 320 pm).

6.1.2 STML Measurements

For the STML measurements, first, the tip was placed on Au(111) with a low sample bias voltage of $\sim$1.5 V. Then the tip was moved on top of the isolated Tpdx-sNDI molecule while keeping the current setpoint low (I = 2 pA). The bias voltage and the setpoint were increased in small steps until an optical spectrum of molecular origin was observed in

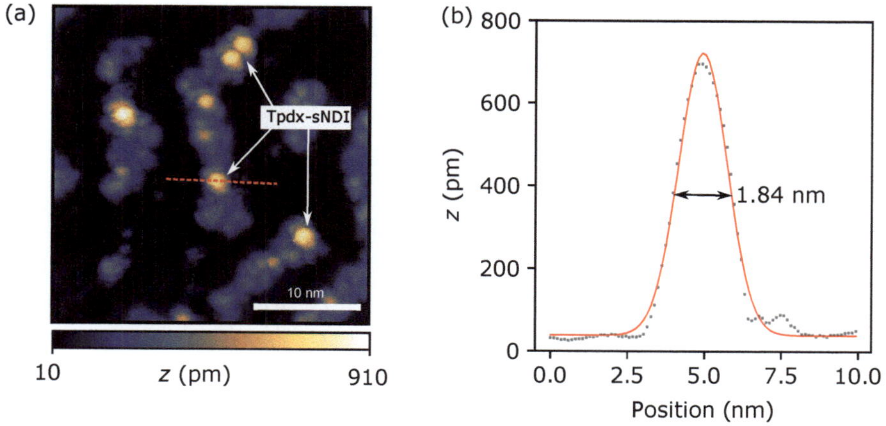

Figure 6.2: Topography and the height profile. **(a)** STM image of Tpdx-sNDI in the as adsorbed state (U = 2 V, I = 2 pA). z is the apparent height. **(b)** Height profile of a single Tpdx-sNDI along the red dotted line marked in **a**.The line profile is fitted with Gaussian peak (red line).

the live view. Figure 6.3 shows a typical optical emission spectrum (in black) recorded on an isolated Tpdx-sNDI molecule at positive bias voltage of $U = 2.62$ V. The spectrum shows a sharp bright main peak (Q, red shaded region) at ~ 2.235 eV (~ 556 nm) followed by a series of low intensity peaks (positions are marked with blue dashed lines) at lower energies. This sharp nature of Q peak indicates an intramolecular optical transition. To analyse the degree of self-decoupling, the photoluminescence (PL) spectrum is plotted together with the STML spectrum as shown in Fig. 6.3 (solid red line) [31]. The PL spectrum has a broad main emission peak ($0 - 0$ line) at 2.233 eV (555 nm) which matches remarkably well with the Q peak of the molecule. This implies that the direct electron transfer between the molecule and the substrate is blocked. As a result, the molecule is sufficiently decoupled from the Au(111) substrate and spends long enough time in the excited state to emit photons into the far-field. Therefore, it is clear that the introduced tripodal platform works effectively to decouple the sNDI and prevents the generation of non-radiative channels. The spectrum in Fig. 6.3 also shows the presence of a broad background which is associated with the plasmonic emission occurring simultaneously.

The peaks of the spectrum in Fig. 6.4a are fitted with Lorentzian peak profile to evaluate the peak positions and their broadening. The width of the Q line is ~ 32 meV which suggests that the environment of the molecule heavily influences the molecular fluorescence. The energy shifts of the vibrational peaks with respect to the emission line Q are shown in table of Fig. 6.4b. Most of the vibrational peaks are part of the functional group region ($1500\ \mathrm{cm}^{-1} < Q - E_n < 4000\ \mathrm{cm}^{-1}$) but some also belong to the fingerprint region as well ($Q - E_n < 1500\ \mathrm{cm}^{-1}$). These vibrational peaks show an irregular distri-

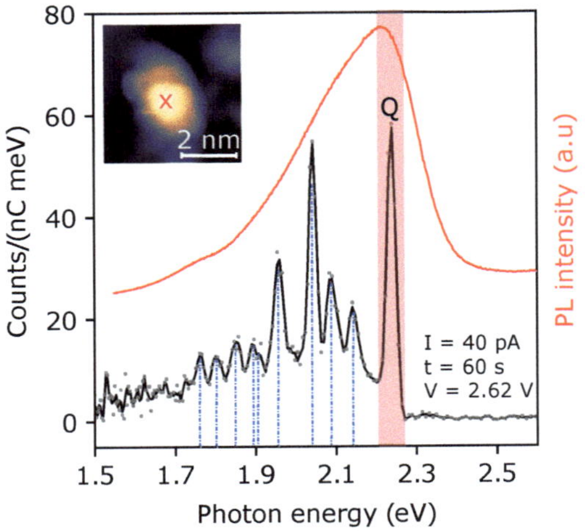

Figure 6.3: STML and photoluminescence spectroscopy. Optical emission spectrum (black line) recorded on an isolated Tpdx-sNDI plotted together with PL spectrum (red line). Inset shows the topography of the molecule with STM tip position marked with a red cross ($U = 2\,\text{V}$, $I = 2\,\text{pA}$). Parameters for recording the spectrum are $U = 2.62\,\text{V}$, $I = 40\,\text{pA}$, $t = 40\,\text{s}$. PL data is provided by Nico Balzer at *Institute of Nanotechnology, KIT.*

bution, which means they belong to different vibrational modes. Later in this chapter, the nature of the vibrational side bands will be discussed in detail for the high resolution data of Tpdx-nNDI, and here the focus is on the nature of the main emission line Q of Tpdx-sNDI.

The spectrum shown in Fig. 6.3 is not unique, in particular, two types of molecules, Tpdx-sNDI(A) and Tpdx-sNDI(B) were identified based on the observed emission spectra. Tpdx-sNDI(A) molecules show unipolar emission with applied bias voltage whereas Tpdx-sNDI(B) molecules show bipolar emission. The optical emission spectrum shown in Fig. 6.3 belongs to Tpdx-sNDI(A).

Tpdx-sNDI(A) Molecules

In Fig. 6.5a, five different spectra are shown at different bias voltages ranging from $-2.84\,\text{V}$ to $2.80\,\text{V}$ on a Tpdx-sNDI(A) molecule. At negative bias voltages, the spectrum does not show any signature of molecular luminescence (in black). As the voltage is increased further to $2.24\,\text{V}$, the Q peak appears followed by several low intensity peaks at lower energies (marked with blue dotted lines in Fig. 6.3). The appearance of the low energy peaks together with Q peak suggests that these are vibrational side bands. In Fig. 6.5b, the integrated intensity of the Q line (red shaded area in fig. 6.5a) is plotted as a function of applied bias voltage, clearly showing that the Q peak has a unipolar

(a)

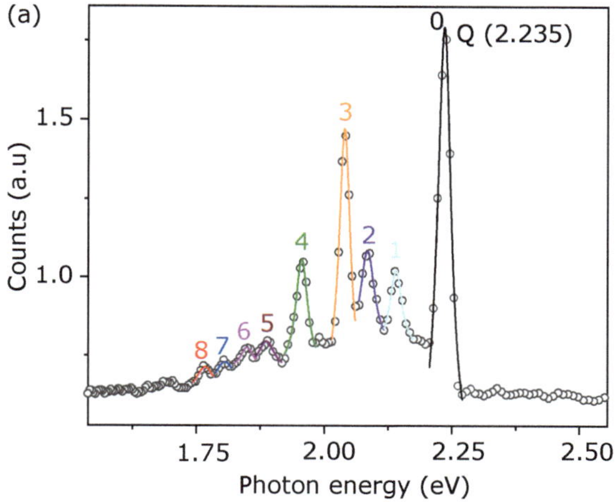

(b)

Peak number (n)	Q-En (meV)	Q-En (cm^{-1})
0	0	0
1	94	758
2	150	1209
3	196	1580
4	280	2258
5	346	2790
6	389	3137
7	433	3492
8	472	3806

Figure 6.4: Vibrational peaks. (a) Optical emission spectrum recorded on an isolated Tpdx-sNDI(A) (U = 2.8 V, I = 40 pA, t = 100 s). Q peak and the vibrational side bands (E_n, n = 1 to 8) are fitted with Lorentzian peak profile (colored solid lines). **(b)** Peak positions of the vibrational side bands (E_n) with respect to the main emission line Q.

emission. Furthermore, the energy of the Q peak ($h\nu_Q$) matches very closely to the onset voltage (eU). This suggests that the molecule is excited by inelastic energy transfer (IET) via plasmons [6, 16] rather than by the recombination of the electron and hole injected from the tip and sample (see chapter 2) [5, 7, 15, 101, 102, 140].

The agreement between the energy of the Q line and the $0-0$ excitation of the PL spectrum (see Fig. 6.3) suggests that the molecule is in its neutral state, which means the excitation via IET should lead to bipolar emission of Q. Instead, an unipolar emission is observed (Fig. 6.5a and b).

This can be attributed to the fact that molecule is directly adsorbed on the metal substrate and the Fermi energy of the tip and the sample are asymmetrically distributed across the molecule. In this case, exciting the molecule in the negative bias voltage may require a much higher energy. However, at higher energies, the molecule is not stable and gets picked up by the tip or disintegrates into the junction and as a result, a sudden increase in the tunnelling current is observed. This observation suggests that the bias dependence of the molecular luminescence is highly dependent on the degree of hybridization of the anchoring group to the metal surface. Such unipolar behaviour has been observed before for other complex molecules, as well as for planer molecules like ZnPc and H_2Pc [7, 102, 141].

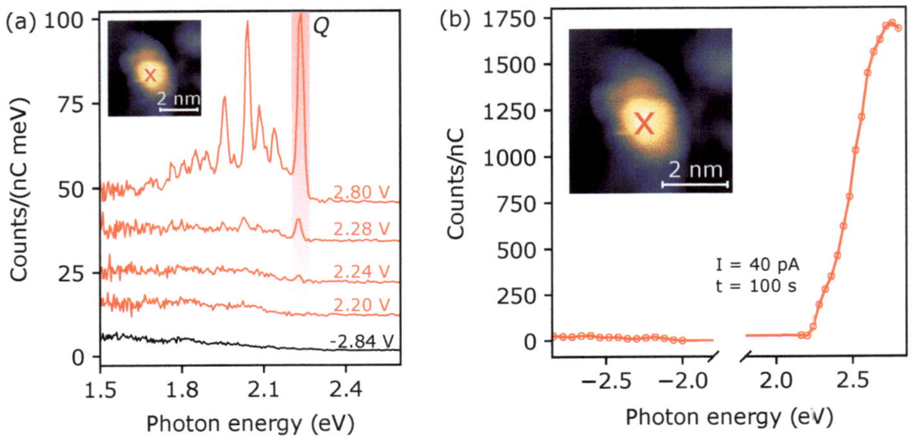

Figure 6.5: Energetics of emission line of Tpdx-sNDI(A). (a) Optical emission spectra recorded at different bias voltages (I = 40 pA, and t = 100 s). **(b)** Integrated photon counts of Q peak in energy rage of 2.198 eV - 2.262 eV. Inset shows the topography of the molecule on which the spectra in **a** are measured, red cross showing the position of the STM tip (U = 2 V, I = 2 pA).

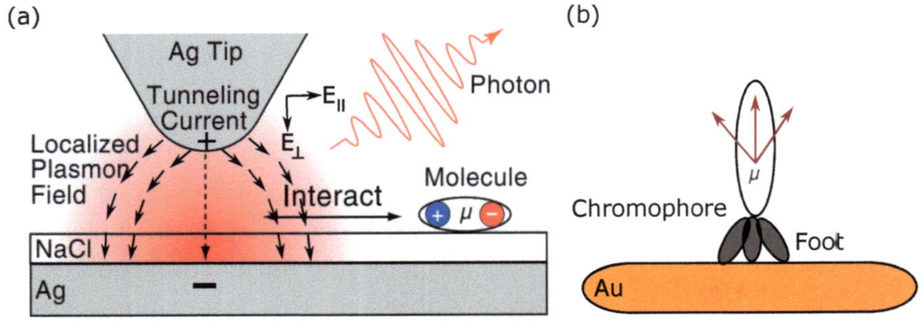

Figure 6.6: (a) Schematic showing the interaction of molecule transition dipole moment (μ) of H_2Pc with the plasmonic field (plasmon-exciton coupling) where μ is oriented parallel to the substrate plane (reprinted with permission from Ref. [21]). **(b)** Schematic showing transition dipole moment (μ) of the Tpdx-sNDI molecule in different orientation with respect to the Au(111) substrate.

Spatial Variation

A molecule in an STM junction is essentially a quantum emitter within a resonant plasmonic nanocavity. The coherent interaction between the discrete states of the molecular emitter and the continuum-like states of the cavity may result in a unique spectral feature called Fano resonance [19, 21, 142–146]. This interaction can affect the energy levels of the quantum emitter and may even shift the transition energy known as photonic Lamb shift [147–149]. In Imada *et al.* [21], the energy dynamics of such a coupled plasmon-exciton system has been discussed for an electronically decoupled single H_2Pc molecule (see Fig. 6.6a). When the tip is placed in the close proximity of the molecule, an asymmetric dip is observed in the expected broad plasmon spectrum, which is called Fano dip. Once the tip is placed on top of the molecule, another channel opens up and electrons can tunnel directly into the molecular orbital, and much enhanced molecular peak is observed. In this particular situation, the transition dipole moment of the molecule (μ) is parallel to the substrate and only the horizontal component of the radially polarized plasmonic field mediate the plasmon-exciton coupling. Therefore it is interesting to see the plasmon-exciton coupling in case of Tpdx-sNDI(A) molecule, where we expect that the molecule has a component of the dipole moment perpendicular to the substrate (see Fig. 6.6a and b) [1].

For this purpose, 10 individual spectra were measured along the the two lines marked in Fig. 6.7a and b (in blue and red, at 90° from each other). Figure 6.7b and c show all individual spectra for the the two line scans. At first glance, it looks like that the molecular emission starts to appear before the tip is on top of the molecule. However, this is not the case as it can be seen from the Fig. 6.7e and f, where the integrated intensity of the Q peak is plotted (in red and blue) together with the relative z positions (in nm) of the STM tip. Importantly, no Fano like behaviour is present.

Therefore, the light emission was only observed once the tip is on the molecule and no plasmon-exciton coupling was observed while the tip is in close proximity of the molecule. This observation can be explained by the fact that in the upright configuration, the transition dipole moment of the Tpdx-sNDI(A) is perpendicular to the substrate, hence the parallel component of the electric field of gap plasmon will have very small effect whereas, the perpendicular component will only strongly couple to the molecular transition dipole when the tip is on the molecule. However, once the tip is on top of the molecule, the perpendicular component couples to the transition dipole moment of a molecule and intense molecular emission is observed and the Fano like line shape is not apparent.

[1]The substitutions are on the sides of the NDI core.

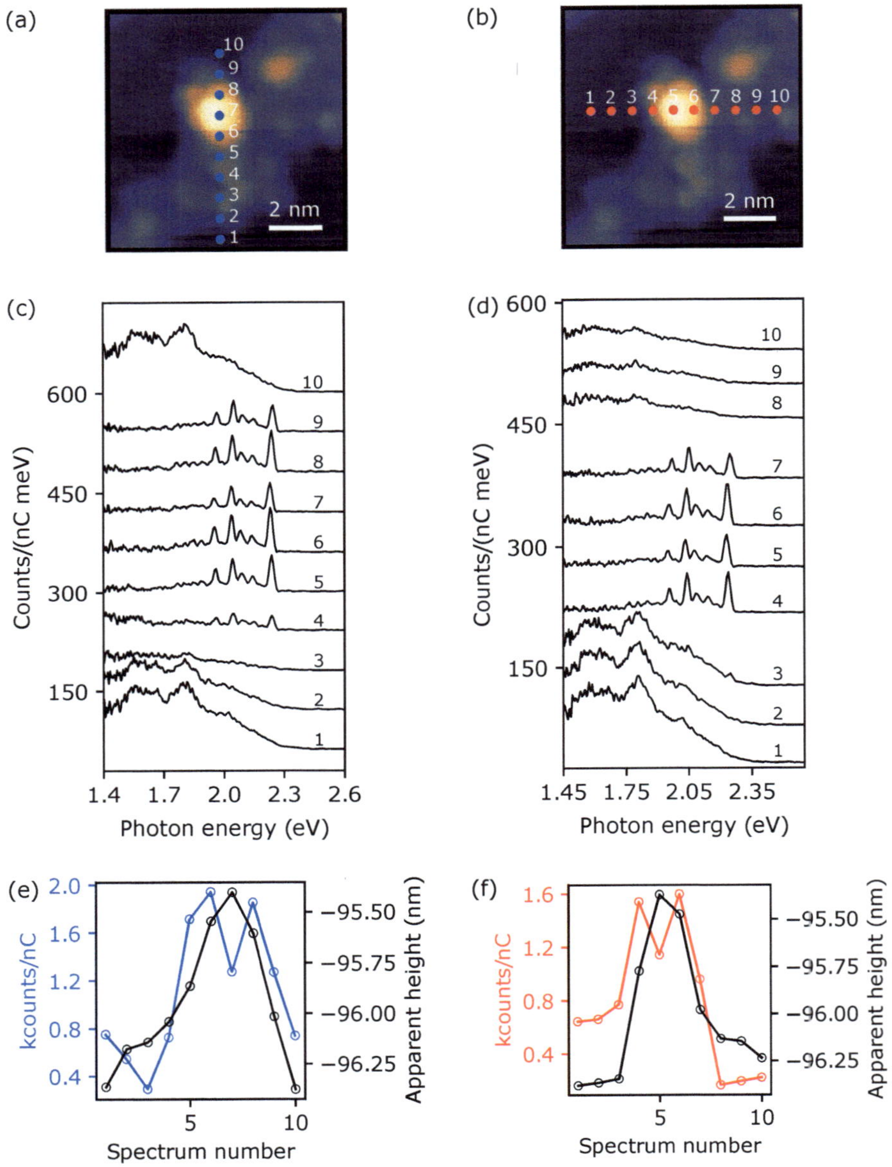

Figure 6.7: (a) and (b) Topographies of a Tpdx-sNDI(A) molecule. Ten tip positions are marked in blue and red where STML spectra were recorded. **(c) and (d)** Series of STML spectra recorded at positions marked in **a and b** respectively. Measurement parameters are $U = 2.6$ V, $I = 30$ pA, $t = 60$ s for both the line scans. **(e) and (f)** Integrated intensity of Q lines (blue and red curve) on the ten points marked in **a and b** with z positions of the tip.

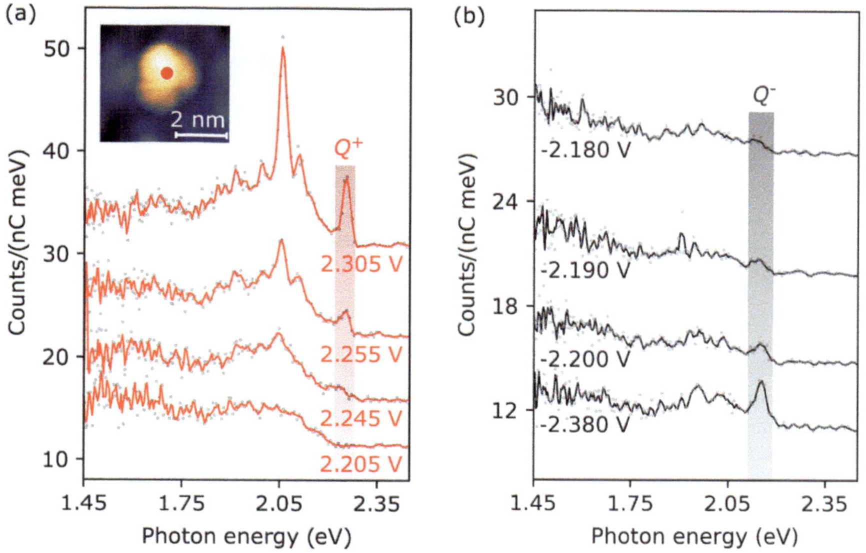

Figure 6.8: Energetics of emission line of Tpdx-sNDI(B). (a) and (b) Optical emission spectra recorded at positive (in red, $I = 100\,\text{pA}$, $t = 60\,\text{s}$) and negative (in black, $I = 100\,\text{pA}$, $t = 100\,\text{s}$) sample bias voltages. Inset in **a** shows the topography of the molecule with the position of the tip marked (red dot).

Tpdx-sNDI(B) Molecules

The second family of the molecules show bipolar emission with applied sample bias voltage. For a better understanding of the bipolar emission of Tpdx-sNDI(B) molecules, several spectra were recorded with different bias voltages ranging from $-2.180\,\text{V}$ to $2.305\,\text{V}$ as shown in Fig. 6.8a and b.

At positive bias voltages (> $2.245\,\text{V}$), a high energy peak (Q^+) is observed at $\sim 2.25\,\text{eV}$ whereas at negative bias voltages (< $-2.15\,\text{V}$), contrary to the spectra measured on the Tpdx-sNDI(A) molecule (Fig. 6.5a), a high energy peak (Q^-) at $\sim 2.15\,\text{eV}$ is observed. For the emission lines Q^+ and Q^-, many peaks at lower energies are observed as well. These low energy peaks only appear once the Q^+ and Q^- peaks are excited, suggesting that these are also vibrational side bands. Similar to the emission from Tpdx-sNDI(A), emission from Tpdx-sNDI(B) happens as soon as energy of the electron (eU) is equal to the energy of the emitted photon ($h\nu$, energy of Q^+ and Q^- peaks).

It is important to mention here that topographically, Tpdx-sNDI(A) and Tpdx-sNDI(B) molecules are indistinguishable.

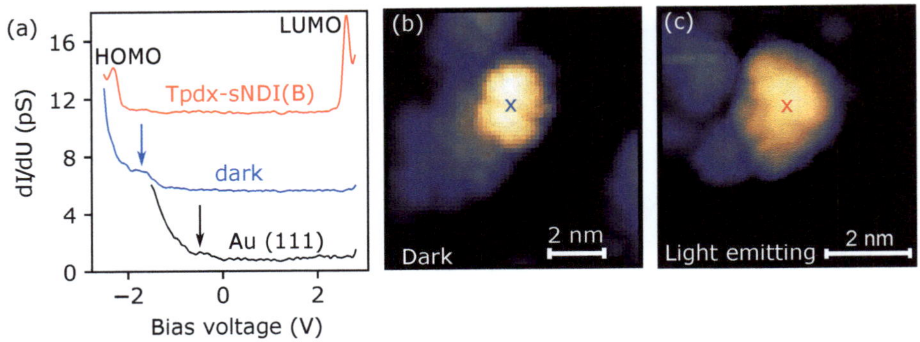

Figure 6.9: Differential conductance measurements. (a) dI/dU spectra measured on Au(111) (black curve, set points $U = -1.5$ V, $I = 400$ pA), on a dark molecule (blue curve, $U = -2.5$ V, $I = 400$ pA), and on light emitting Tpdx-sNDI molecule (red curve, $U = -2.5$ V, $I = 380$ pA). **(b) and (c)** Topographies of a dark and a light emitting Tpdx-sNDI(B) molecule with the positions marked where dI/dU measurements were performed (blue and red cross).

Electronic Structure of Tpdx-sNDI(B)

For a better understanding of the electronic structure of different configurations of the molecules, systematic differential conductance (dI/dU) measurements were carried out on a Tpdx-sNDI(B) molecule as well as on a molecule which did not show molecular luminescence (dark molecules). Figure. 6.9a shows three dI/dU spectra, on Au(111) (in black), on a dark molecule (in blue), and on a light emitting molecule (in red) while keeping the STM tip unchanged (topography in Fig. 6.9b and c). Tpdx-sNDI(A) molecules were not very stable in the junction and the attempts to perform the dI/dU measurements were unsuccessful.

The spectrum on Au(111) does not show any sharp feature except the surface state at -0.48 V (black arrow). This ensures that the variations in the differential conductance measurements on the molecules come from the different electronic state of the molecules. On a dark molecule, a broad peak at around -1.70 V is observed (blue arrow) whereas on a light emitting molecule (Tpdx-sNDI(B)) a sharp peak at -2.31 V (red spectrum) is observed. At positive bias, only the light emitting molecule shows a sharp peak centered around 2.61 V. This observation goes along the argument of certain electronic decoupling of the corresponding orbitals, which are tentatively assigned as HOMO (-2.31 V) and LUMO (2.61 V) of the light emitting molecule. Moreover, light emission from the molecule is only observed when both the involved orbitals are decoupled as evident from the appearance of LUMO. Here, the difference between the HOMO-LUMO from the dI/dU measurements does not corresponds to the optical gap observed from the light emission spectra (optical gap is equal to the photon energy of

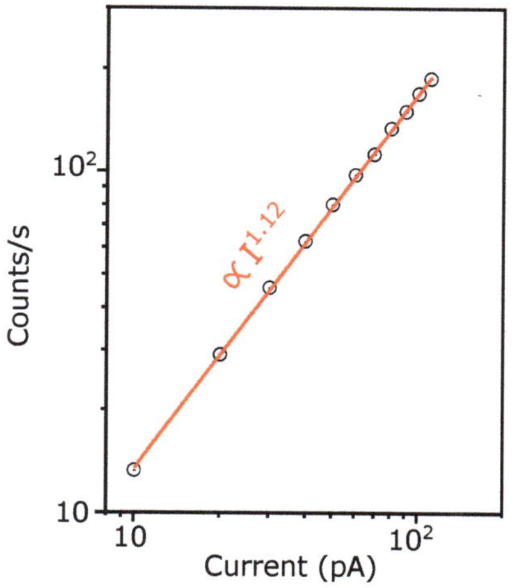

Figure 6.10: Intensity of Q peak as a function of tunnelling current (I). Measurement parameters are U = 2.59 V, t = 60 s.

the Q peak for Tpdx-sNDI(A), and to Q^+/Q^- for Tpdx-sNDI(B)), as STM measure the energy of the states including the charging energy.

The underlying mechanism for the light emission from the Tpdx-sNDI molecules is based on the IET model [6, 16] discussed in chapter 2. The adaptation of the IET model is due to the fact that the energies required to excite the three emission lines (Q, Q^+ and Q^-) match very closely (less than few meV) to the energies of the emission lines. Although the model explains the emission of the three lines, it is, however, not complete to explain the differences observed in the efficiencies and their dependencies on the polarity of bias voltage.

Conclusion of the Experiments on Tpdx-sNDI

From the STM/STML measurements on Tpdx-sNDI the following conclusions can be made:

- Based on the light emission spectra, two kinds of Tpdx-sNDI molecules are observed, unipolar Tpdx-sNDI(A) molecules with main emission peak Q, centered at 2.235 eV at positive bias voltage, bipolar Tpdx-sNDI(B) molecules with main emission peaks Q^+ and Q^-, centered at 2.25 eV and 2.15 eV at positive and negative bias voltages, respectively.
- The observed quantum efficiency of the Q peak of Tpdx-sNDI(A) is in the order of $\sim 10^{-6}$ photons/electron.

- The observed quantum efficiencies of the Q^+ and Q^- peaks of Tpdx-sNDI(B) are in the order of $\sim 10^{-7}$ and $\sim 10^{-8}$ photons/electron, respectively.
- Bias dependence measurements suggest that the light emission from the molecules can be explained by IET model. Moreover, the linear dependency of the intensity of Q peak on tunnelling current suggest one-electron process (see Fig. 6.10).
- Differential conductance measurements show that the light emitting molecules are decoupled from the substrate. From Fig. 6.9a, it can be said that both the orbitals involved in the light emission need to be decoupled.
- Spatial dependence of the luminescence spectra showed no sign of Fano-line shape when the tip is in close proximity of the molecule.

6.2 Experiments on Tpdx-nNDI

Optical emission spectra of Tpdx-sNDI establish that the proposed tripodal platform works well to decouple the light emitting chromophore from the metal substrate. This is where the superiority of this chemical route of decoupling lies. It enables us to function-alize different light emitting chromophores on the tripodal platform and hence tune the energy of the emission lines. To further test the reproducibility of this proposed tripodal platform, another chromophore -nNDI is anchored to it with the expected high light emission quantum efficiency [31]. In the following sections, a detailed STM/STML study will be shown on a single self-decoupled Tpdx-nNDI molecules.

6.2.1 Sample Preparation and Topography

Deposition of Tpdx-nNDI on a clean Au(111) surface was done with a similar spray deposition method as for Tpdx-sNDI discussed in section 6.1.1. After the deposition, the samples were annealed at 100 °C and subsequently placed in the STM. Figure 6.11a shows the typical STM image of the sample, where each isolated high object is assumed to be single Tpdx-nNDI molecule. These molecules show apparent height and width of $\sim$700 pm and $\sim$1.94 nm (FWHM), respectively. As discussed before (section 6.1.2), real height in STM is not accessible directly whereas a much closer estimate can be made with for the width of the molecule. In case of the Tpdx-nNDI molecule, the calculated width (MM2 minimization in ChemOffice) is $\sim$1.76 nm and agrees well with the measured width of the molecule.

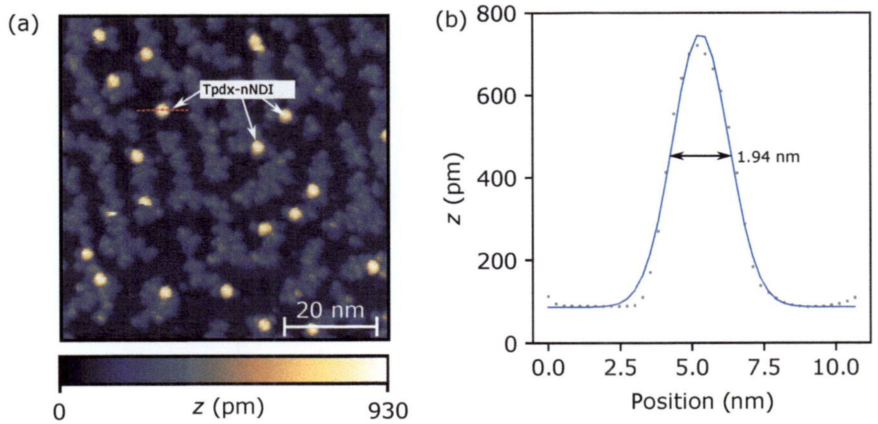

Figure 6.11: Topography and the height profile. (a) STM image of Tpdx-nNDI in the as adsorbed state (U = 1.5 V, I = 2 pA). **(b)** Height profile of a single Tpdx-nNDI along the line (red dotted) marked in **a**. Blue line is the Guassian fit with the FWHM of 1.94 nm.

6.2.2 STML Measurements

To record the STML spectrum of Tpdx-nNDI, a similar procedure was followed as for the Tpdx-sNDI. First, the STM tip was placed on top of an isolated Tpdx-nNDI while keeping low sample bias voltage (~1.8 V) and low setpoint current (~2 pA). Then the bias and the setpoint current were increased gradually until a molecular luminescence was observed. Figure 6.12 shows a typical STML spectrum (in black) recorded on an isolated Tpdx-nNDI molecule. It is plotted together with the PL spectrum of the molecule measured in solution [31]. The STML spectrum in Fig. 6.12 shows two main sharp peaks at ~ 1.974 eV (Q_1) and ~ 2.231 eV (Q_2).

It is important to mention that the peak positions of the emission lines of Tpdx-sNDI and Tpdx-nDNI are not fixed and sensitive to the applied bias voltage as well as to the measurement positions on the molecules. A detailed discussion of the change in the peak position will be presented later in the chapter (see section 6.3).

The sharp emission lines (Fig. 6.12) indicate an intramolecular optical transition as their origin. The PL spectrum shows a main peak $(0-0)$ at ~ 1.98 eV (~ 625 nm) and a less prominent shoulder peak at ~ 2.21 eV (~ 560 nm). The agreement between the STML peaks (Q_1/Q_2) and the PL peaks suggests that the chromophore unit is sufficiently electronically decoupled from the electron bath of the substrate, specially the energy of Q_1 peak matches remarkably well with the main $0-0$ PL peak. For such self-decoupled single molecule, a record high emission efficiency ($\sim 1.1 \times 10^{-3}$ photons/electron) was observed (see section 6.3.1).

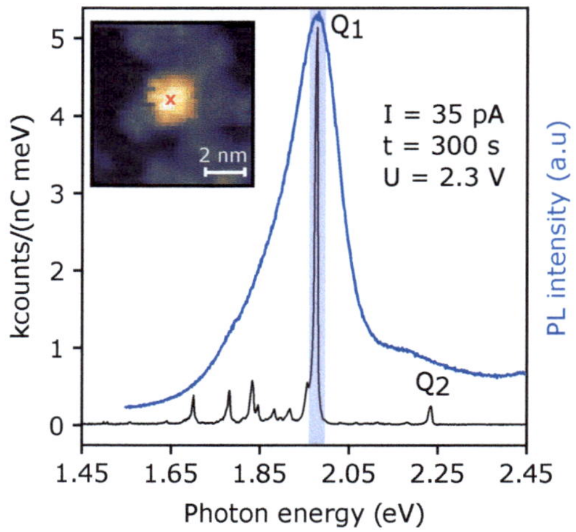

Figure 6.12: STML and photoluminescence spectroscopy. STML spectrum (in black) recorded on an isolated single Tpdx-nNDI. Inset shows the topography of the molecule with STM tip position marked with a red cross ($U = 1.5\,\mathrm{V}$, $I = 2\,\mathrm{pA}$). PL data is provided by Nico Balzer at *Institute of Nanotechnology*, KIT.

Here, the Q_1 peak is much sharper compared to the Q peak of the Tpdx-sNDI molecule because the measurement was carried out with higher spectrometer resolution. The width of the Q_1 peak is $\sim 7\,\mathrm{meV}$ whereas the Q_2 peak has a width of $\sim 9\,\mathrm{meV}$. This indicates that chromophore is fully decoupled and unaffected by the presence of nearby electrodes. In this case, the linewidth of the emission lines are limited by other factors, such as spectrometer resolution, interactions with phonons and plasmons. The lifetime of the Q_1 peak is 0.047 ps for the linewidth of 7 meV which is still much shorter than the expected fluorescence lifetime.

6.2.3 Energetics and Bias Spectroscopy

Similar to the Tpdx-sNDI, the molecular emission of Tpdx-nNDI is also not unique and different kinds of emissions were observed. Tpdx-nNDI can be categorized as Tpdx-nNDI(A) and Tpdx-nNDI(B) molecules based on their emissions (see Fig. 6.13a-d). In the previous section, the discussed emission was from Tpdx-nNDI(A) molecule (see Fig. 6.12). For Tpdx-nNDI(B) molecules, only one main peak at $2.253\,\mathrm{eV}$ (Q^+) is observed at positive bias (see Fig. 6.13b). Both types of emissions are bipolar in nature.

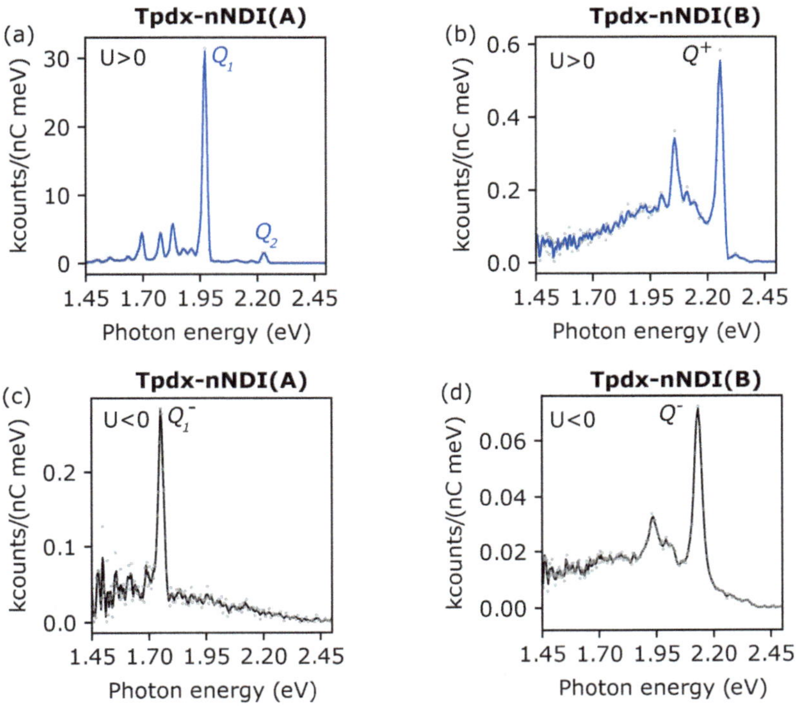

Figure 6.13: Molecules with different emissions. **(a)** Emission from Tpdx-nNDI(A) at positive sample bias voltage with two main emission peaks Q_1 and Q_2 ($U = 2.40$ V, $I = 15$ pA, $t = 20$ s). **(b)** Emission from Tpdx-nNDI(B) at positive sample bias voltage with one main emission peak Q^+ ($U = 2.35$ V, $I = 30$ pA, $t = 10$ s). **(c) and (d)** Emission from Tpdx-nNDI(A) and Tpdx-nNDI(B) molecules at negative sample bias voltages with main emission peaks Q_1^- ($U = -2.35$ V, $I = 15$ pA, $t = 20$ s) and Q^- ($U = -2.35$ V, $I = 30$ pA, $t = 60$ s), respectively.

Tpdx-nNDI(A) Molecules

Bias Dependence

For further understanding of the emission from Tpdx-nNDI(A), evolution of the optical spectrum as a function of the applied bias was recorded for both polarities (see Fig. 6.14a and b). In Fig. 6.14a, three spectra at positive biases are shown. At 1.95 V, the Q_1 peak starts to appear (marked in red circle) and as soon as the bias voltage is 1.97 V, a fully developed Q_1 peak appears together with the low energy vibrational side bands. The energy of Q_1 peak is 1.961 eV. Similar to Tpdx-sNDI, the onset voltage energy matches the energy of the Q_1 peak ($eU = h\nu_{Q_1}$). A similar behaviour is observed for the Q_2 peak as well ($eU = h\nu_{Q_2}$). Figure 6.14b shows spectra recorded at negative bias voltages. At negative bias, only one peak (Q_1^-) is observed at much lower energy (1.751 eV). This

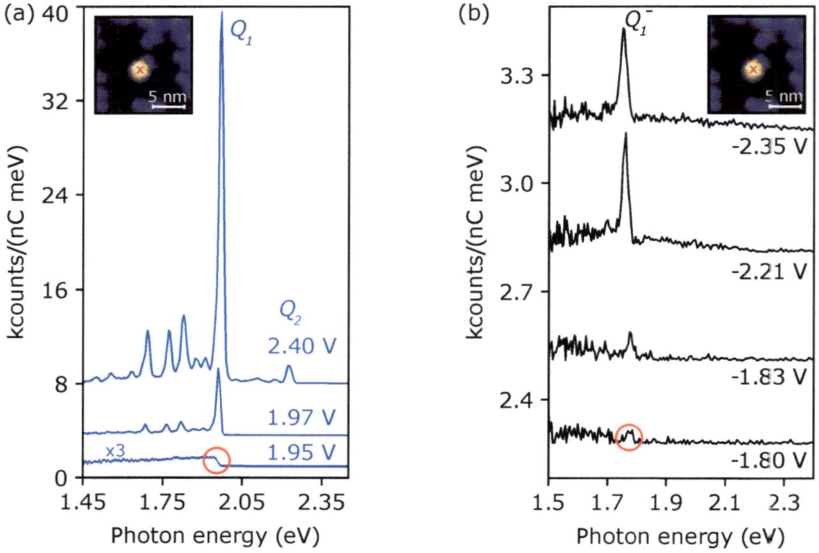

Figure 6.14: Bipolar emission of Tpdx-nNDI(A). (a) and (b) Optical emission spectra recorded at different positive and negative sample bias voltages (I = 15 pA, t = 20 s). Insets show the topographies of the molecules with tip positions marked (red cross), U = 1.5 V, I = 2 pA.

shift in the energy could be the signature of Stark effect, however, this is a large shift compared to the previous reports on STML measurements on single molecules. This observation can be attributed to the fact that molecular dipole has a large component parallel to the electric field. In negative bias, the emission of Q_1^- peak only starts to appear once the onset energy $eU = h\nu_{Q_1^-}$. These observations point toward the fact that emission from the Tpdx-nNDI(A) is mediated by IET via plasmons (see chapter 2). A major difference between the spectrum at different polarity of bias voltages is in their emission quantum efficiencies. At positive bias, the emission efficiency is in the range of $\sim 10^{-4} - 10^{-3}$ photons/electron, whereas at negative bias the efficiency is 2-3 order less and is in the range of $\sim 10^{-6} - 10^{-7}$ photons/electron. The origin of this low emission efficiency is unclear at this point.

Up-conversion Electroluminescence (UCEL)

For Tpdx-nNDI, another very interesting phenomenon was observed, known as up-conversion electroluminescence (UCEL). In UCEL the emitted photon has a higher energy than that of the excitation electron ($eU < h\nu$). This has also been observed in other molecules in other STML experiments [6, 63, 150–155]. However, for single self-decoupled molecules, this has not been observed, possibly due to the lack of junc-

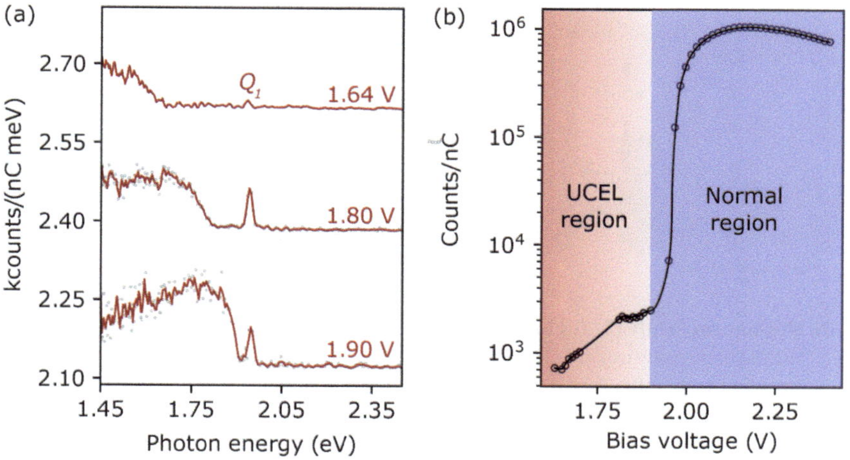

Figure 6.15: UCEL of Q_1 peak of a Tpdx-nNDI(A). (a) Optical emission spectra recorded at bias voltages lower than the threshold voltage required for the excitation of Q_1 peak (I = 10 pA, t = 30 s). **(b)** Integrated intensity of Q_1 peak (1.931 eV - 1.996 eV) plotted as function of sample bias voltage showing two distinct regions for UCEL (shaded red) and Normal (shaded blue) emissions.

tion stability. Figure 6.15a shows spectra recorded at low bias (< 1.96 V) (UCEL region). Furthermore, a distinct UCEL feature can be observed until as low bias voltage as 1.64 V, below this bias voltage, emission is not observed. This could be attributed to the extremely low emission efficiency at such low voltages. This phenomenon can be explained by the spin-triplet-mediated up-conversion model proposed by Chen *et al.* [24] while excluding the higher order tunnelling mechanism due to the measurements at low tunnelling current, see Fig. 6.16. For higher order tunnelling mechanism UCEL would already occur at bias voltage of 0.98 V which is the half of the molecular optical gap. The model given in the Fig. 6.16 explains the emission from H_2Pc molecule, decoupled with NaCl layers at negative bias voltages .

Therefore, more than one mechanism is involved in the light emission of Q_1 peak at different bias voltage regions. When the energy of the excitation electrons is equal or higher than the peak energy of the emission line ($eU \geq h\nu_{Q_1}$), then the most likely mechanism is mediated by IET, whereas as soon as the energy of the excitation electron is less than the peak energy of emission line ($eU < h\nu_{Q_1}$), UCEL occurs and a different mechanism occurs. This difference can be seen very clearly in Fig. 6.15b where the integrated intensity of the emission line Q_1 is plotted as a function of the bias voltage. The intensity of the Q_1 peak shows two distinct stepwise increments in different voltage regions (marked with shaded brown and blue), implying that two different processes

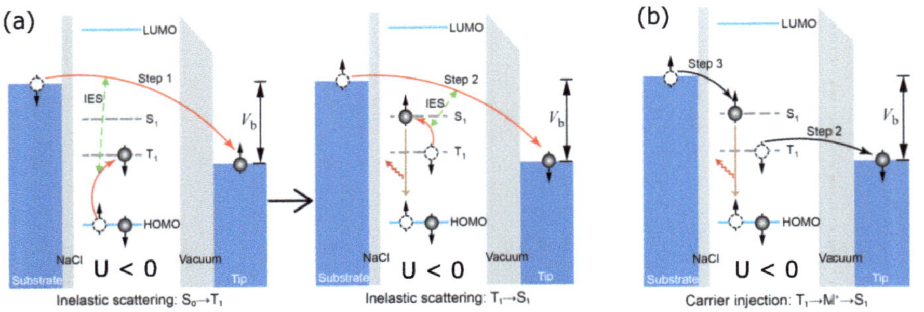

Figure 6.16: Mechanism for UCEL. (a) In the UCEL region ($eU < h\nu_{Q_1}$), an electron is excited from the HOMO to the spin-triplet (T_1) state by IET via plasmon (left). Hereafter one possibility of photon emission is that the electron in T_1 is again excited via another plasmon to S_1 state and then this electron de-excites by emitting a photon into far-field (right). **(b)** Another possibility is that electron stays long enough in the T_1 state to tunnel into the tip allowing another electron to tunnel from substrate to S_1 state. This electron in S_1 then de-excites and emits a photon into far-field (reprinted with permission from Ref. Chen *et al.* [24]).

prevail in the two voltage regimes. This observation is consistent with the measurements done on single H_2Pc molecules decoupled by NaCl layers [24].

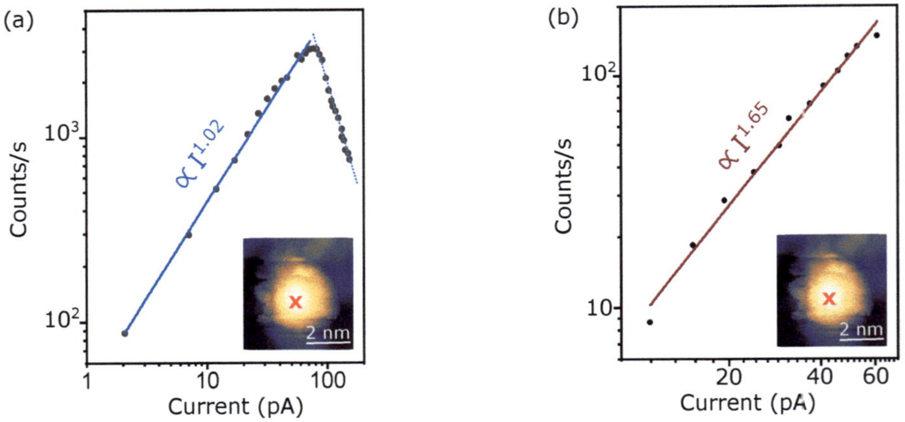

Figure 6.17: Intensity of Q_1 line as a function of tunnelling current. (a) Intensity of Q_1 as a function of tunnelling current (I) in the normal emission range (U = 2.4 V). Inset shows the topography of the molecule and the tip position is marked with a red cross (U = 1.5 V, I = 2 pA). **(b)** Intensity of the Q_1 peak as a function of tunnelling current (I) in the UCEL range (U = 1.88 V). The inset shows the topography of the molecule and the tip position is marked with a red cross (U = 1.5 V, I = 2 pA).

The two regions can further be distinguished by the current dependent measurements. Figure 6.17a and b show integrated intensity of Q_1 peak as a function of the tunnelling current in normal emission and in UCEL regions, respectively. In the normal emission region, the intensity of the Q_1 line scales linearly ($\propto I^{1.02}$) with the current till 72 pA, however, above this value, the photon counts decreases with steeper slope ($\propto I^{2.20}$) (see Fig. 6.17a). This observation is in contrast to the current dependent measurements done on planer molecules like ZnPc or H_2Pc [23, 29] and can be explained by the fact that Tpdx-nNDI(A)s are standing upright and the tip is much closer to the Tpdx-nNDI(A) compared to the planer molecules. As the current increases above 72 pA, the tip comes even closer to the Tpdx-nNDI(A) molecule and due to the increased overlap between the wavefunctions of the tip and the chromophore, non radiative decay becomes possible. Moreover, this creates the same situation as the molecules on metal surface and as a results, fluorescence is quenched. In the UCEL region, the intensity of Q line increases non-linearly as shown in Fig. 6.17b ($\propto I^{1.65}$) as expected in a multielectron process.

Analysis of Vibrational Side Bands

It is clear from the optical emission spectrum of Tpdx-nNDI(A) that there are several low intensity peaks at lower energies than the Q_1 line (see Fig. 6.12). These low energy (LE) peaks are assigned to the here-called vibrational LE-band [16, 141] (see Fig. 6.18). Figure 6.19a shows the energy diagram of an electronic transition from the electronic ground state (S_0) to the first excited state S_1 together with the associated LE-band vibrational transitions for a single mode. In this energy diagram, Kasha's rule is followed which states that the system should relax to the ground vibrational state of S_1 before decaying to the ground vibrational state of S_0 [156].

Above the energy of Q_1 peak, it seems like there is only one more high energy peak (Q_2), however, this is not the case. Once the spectrum is enlarged, it is clear that there is a series of high energy peaks present between Q_1 and Q_2 emission lines, see Fig. 6.18. Here, these peaks are called high energy band (HE-band). It is very interesting to notice that these HE-band peaks are mirror symmetric to LE-band peaks with respect to Q_1 emission line. Due to this symmetry, it can be interpreted that these HE-band peaks are anti-Stokes lines. However, this is not true for two reasons: First, molecules are at 4.4 K, so there are no thermally excited vibrational states of S_0 from which an anti-Stokes line can originate. Secondly, the bias dependent measurements in Fig. 6.20a clearly show that these HE-band peaks appear as the bias corresponding to their energy is applied. However, anti-Stokes lines should be apparent as soon as the Q_1 peak is excited. Moreover, bias dependent measurements also suggest that they are not LE-band of a different electronic transition at higher energy (i.e, Q_2 peak). Therefore, based on these observations, it can be deduced that the HE-band peaks originate from a direct transition between the vibrationally excited state of S_1 to the ground vibrational state of S_0, although this transition will be the breaking of Kasha's rule. In recent STML studies,

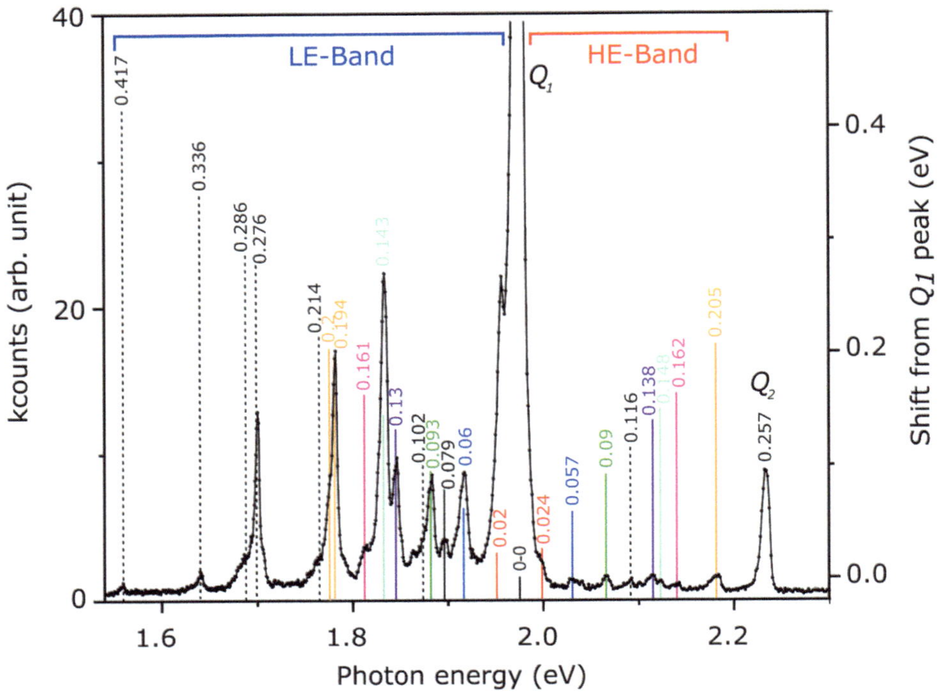

Figure 6.18: Optical emission spectrum of Tpdx-nNDI(A). Spectrum is zoomed in y-axis (right) to highlight the low intensity LE-band and HE-band peaks. Q_1 line is marked with solid black line. On the right (left) of the Q_1 line, all the LE-band (HE-band) peaks are marked with colored (black), solid (dotted) lines. Color of left and right vibrational lines are same if their shift from Q_1 line is almost same. LE-band peaks which are marked with dotted black lines do not have their mirror symmetric peak in HE-band. Height of each solid(dotted) lines represent their shift from Q_1 peak (on right y-axis) also written on the lines.

such type of transition has been observed and it is called hot luminescence (HL) [6, 157, 158]. An energy diagram for such transitions is represented in Fig. 6.19b for a single vibrational mode.

Energy diagrams in Fig. 6.19a and b represent the harmonic progression of a single vibrational mode similar to the previously reported STML studies on single molecules [5, 101, 115, 140]. In these reports, the intensity of the peaks depends on the so-called Franck-Condon factors, and their dispersion is linear under harmonic approximation. In contrast to this, the vibrational bands of Tpdx-nNDI(A) have irregular peak dispersion (see Fig. 6.19c where shift of LE-band and HE-band from Q_1 emission line are plotted). This suggests that the peaks in both side bands are from different vibrational modes and probably only the transition of the first harmonic of each mode has enough intensity to

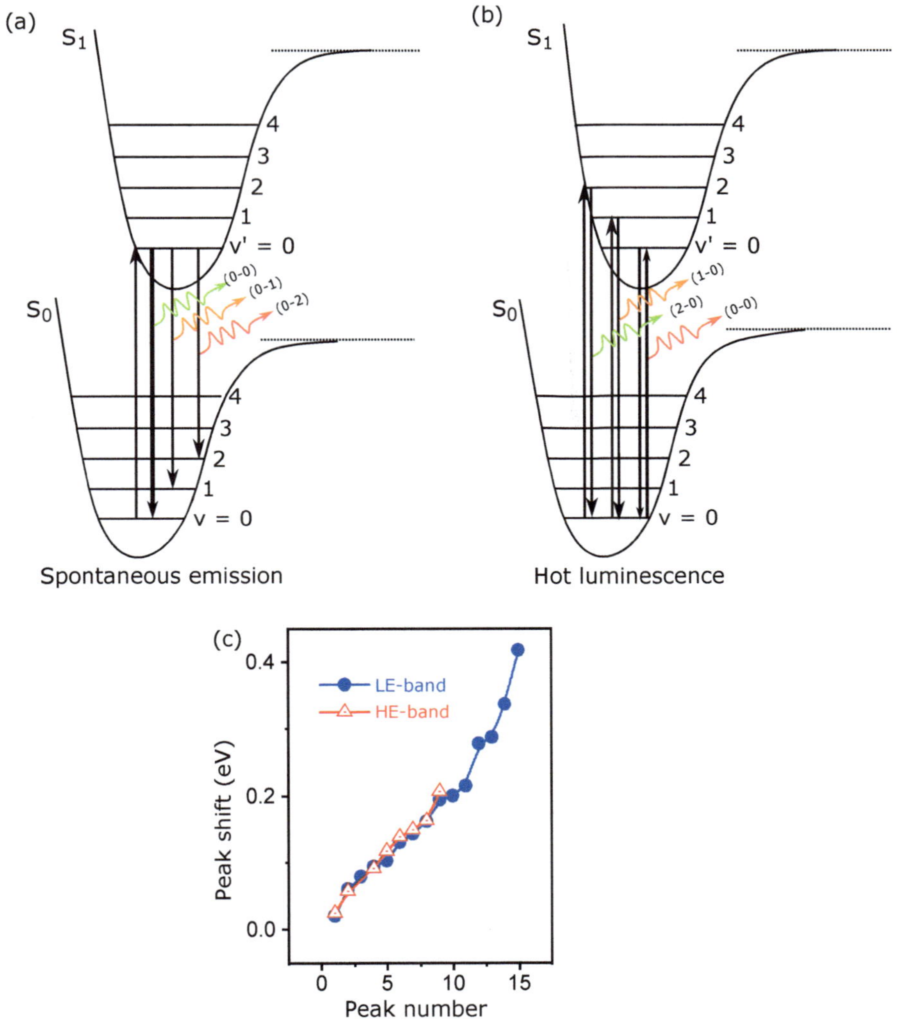

Figure 6.19: Origin of LE-band and HE-band. (a) Energy diagram of an electronic transition from ground state S_0 to excited state S_1. In this diagram Kasha's rule is followed which means molecule is first excited into an excited vibrational state of S_1 and then due to high vibrational damping rate, the molecule first relaxes to the ground vibrational state of S_1 before making a transition to the ground state S_0 and emitting light into the far-field. **(b)** Energy diagram representing the transition leading to HL by radiatively de-exciting directly from the excited vibrational states of S_1 to the vibrational ground state of S_0. **(c)** Shift of LE-band and HE-band peaks from the Q_1 emission line plotted with peak number. The plot clearly shows the irregular dispersion of the vibrational modes.

be detected. A comprehensive picture of this situation is shown in the Fig. 6.20b, where a molecule is present between two metallic leads (in our case tip and sample).

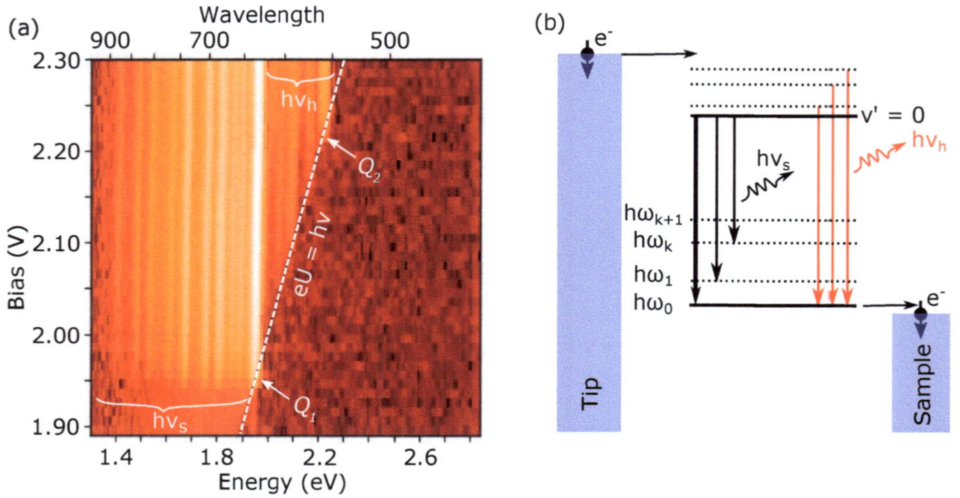

Figure 6.20: Evolution of LE-band and HE-band peaks. Color map of optical emission spectra recorded at different sample bias voltages. White dotted line represents the $eU = h\nu$ condition. **(b)** A comprehensive picture representing transitions leading to emission of the LE-band peaks ($h\nu_s$) and HE-band peaks ($h\nu_h$) from a molecule between two metallic leads (tip and sample in our case). $h\omega_k$ represents the first harmonic of different vibrational modes.

In a typical STML experiment, only the LE-band is observed together with the $0-0$ transition. The presence of the LE-band alone provides the spectroscopic fingerprint of the molecular species. However, since the peaks of LE-band are transitions from the vibrational ground state of S_1 to the different vibrational state of S_0, only the mapping of S_0 vibrational level is possible. The optical emission spectrum of Tpdx-nNDI, however, also shows the HE-band and thus provides information about the vibrational levels of S_1 state. Here, it is very interesting to observe that by exploiting the localized plasmon resonances in the STM cavity, vibrational fingerprints of molecules can be achieved, which is typically obtained by Raman spectroscopy or in AFM/STM community by tip-enhanced Raman spectroscopy (TERS) [159–163]. This would be the first time that such spectroscopic and spatial resolution has been observed for a well-isolated self-decoupled single molecule.

Further theoretical studies of Tpdx-nNDI(A) molecules can provide an insight into the involvement of different modes which appear in the HE-band. Furthermore, the calculated Raman modes of the molecules and experimental Raman spectroscopic data could also give more information about the transitions leading to LE-band and HE-band peaks for the particular molecule.

Possible Mechanisms for HL

Based on the experimental observations, the first thing that can be said about the relaxationless HL of the molecules is that this is in contradiction with the Kasha's rule. The theoretical studies to understand the exact nature and the origin of these HL peaks are beyond the scope of this work, but it is important to discuss the previous theoretical models proposed to explain the HL [137, 145, 164–168]. After the first STML observation of HL by Dong *et al.* [6], one of the first theoretical treatment was done by Seldenthuis *et al.* [164] where the effect of the plasmons were ignored. They stated that HL can be observed as soon as the vibrational relaxation rate becomes comparable to the electron transmission rate ($\sim 10^{10}\ s^{-1}$). Followed by this, Tian *et al.* [137] and Zhang *et al.* [166] took into account the effect of plasmons. Zhang *et al.* [166] considered a system of a molecule sandwiched between two metal nanoparticles, whereas Tian *et al.* [137, 165] considered a more closer model to the situation in STM junction. In the later model, they mimic the plasmon as a sequence of pulses because the plasmon is generated by inelastic electron tunnelling. In their simulation, they see the evolution of HL as soon as they consider the inclusion of both electron tunneling and plasmonic excitation.

Followed by the work of Tian *et al.* [137, 165], another theoretical study was reported by the same group who first observed the HL in STM [168]. They proposed a different mechanism for HL by arguing that the plasmonic field excited directly by inelastic tunneling electrons is probably very weak after insertion of the molecule and it is complex to access the extent of plasmon exciting the molecule in return. In their model, they show that once the molecule is placed in a plasmonic nanocavity, the spontaneous decay rate is dramatically enhanced by several orders of magnitude and the molecule may go through the radiative decay before relaxing down to the vibrational ground state of S_1. They also see the effect of the dipole of the molecule and the tip-sample distance on HL. Although after this study, it has been reported in many STML studies of single molecules that plasmons can effectively excite the molecule, including the observations on -sNDI and -nNDI. These theoretical studies explain HL to some extent, but it is clear that full understanding is still lacking and further intense theoretical studies need to be done.

Tpdx-nNDI(B) Molecules

Tpdx-sNDI molecules show either one or the other type of emission (see Section 6.1.2). In the case of Tpdx-nNDI, the molecules may switch from Tpdx-nDNI(A) to Tpdx-nDNI(B). It was observed that in the pristine state, the molecules were in Tpdx-nDNI(A) state. In the presence of the STM tip, the molecules also switch to Tpdx-nDNI(B) state. It is

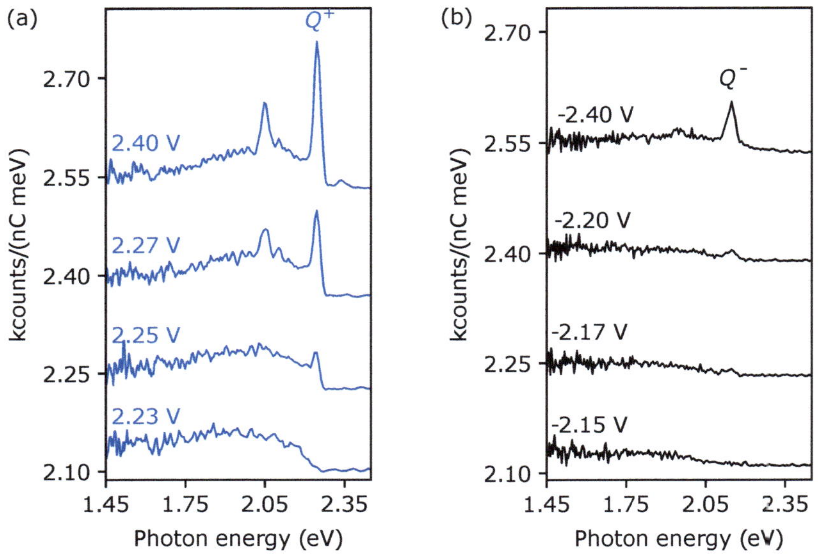

Figure 6.21: Bipolar emission of Tpdx-nNDI(B). (a) Optical emission spectra recorded at different positive sample bias voltages (I = 30 pA, t = 15 s). **(b)** Optical emission spectra recorded at different negative bias voltages (I = 30 pA, t = 30 s).

important to mention here that this switching was very subtle and no obvious change in the apparent height is observed before and after the switching.

Figure 6.21a and b show the evolution of molecular emission spectra as a function of the applied positive and negative sample bias voltages, respectively. At positive bias, no molecular emission was observed till 2.23 V and as the bias is increased further, a sharp peak (Q^+) at ~ 2.24 eV is observed. This Q^+ peak starts to appear as soon as the energy of the exciting electron is equal to energy of the emitted photon ($eU = h\nu_{Q^+}$, IET via plasmon). For Tpdx-nNDI(B), only one main peak Q^+ is observed with energy very close to the Q_2 peak of Tpdx-nNDI(A). At this point, it is not clear whether after switching from Tpdx-nNDI(A) to Tpdx-nNDI(B), only the Q_2 peak remains with slightly different energy or Q^+ has a completely different origin with energy very close Q_2. Therefore, for the moment, it is vital to consider Q^+ as different emission than Q_2. Furthermore, in the negative bias voltage, the molecular peak (Q^-) appears at lower energy ~ 2.142 eV and this small shift can again be attributed to the Stark effect.

Figure 6.22a shows the normalized counts of Q^+ peak in the energy range of 2.186 eV - 2.287 eV as a function of applied positive bias voltage. Here, the emission increases with the bias voltage and has maxima at ~ 2.125 V and further increase in the bias shows

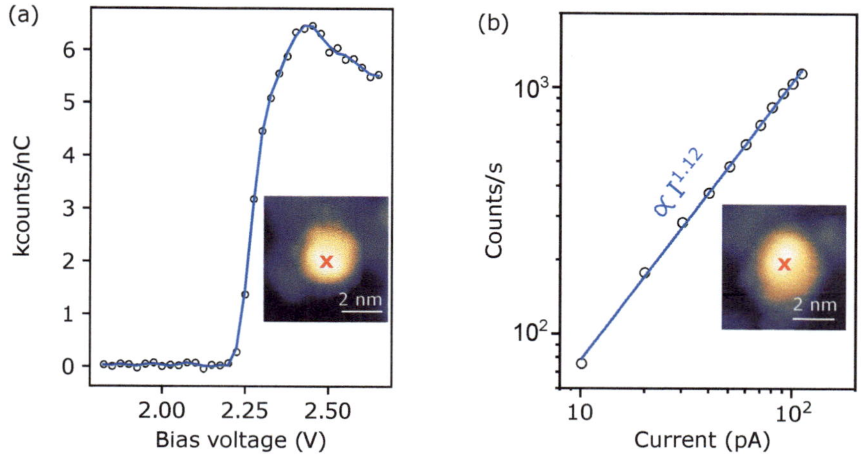

Figure 6.22: Intensity of Q^+ emission line as a function of applied bias and tunnelling current. **(a)** Integrated intensity of Q^+ emission line in the energy range of 2.186 eV - 2.287 eV (I = 50 pA, t = 20 s). **(b)** Integrated intensity of emission line as a function of tunnelling current in the energy range of 2.186 eV - 2.287 eV (U = 2.40 V, t = 15 s). Insets in **a** and **b** show the topography of the measured molecule with tip position marked (red cross) (U = 1.8 V, I = 2 pA).

a step in the intensity. Current dependent measurement of the intensity of Q^+ peak shows a linear dependence, hence pointing towards one electron process, see Fig. 6.22b.

Differential Conductance Measurement on Tpdx-nNDI

Tpdx-nNDI(B) molecules were very stable and it was possible to measure dI/dU to get more information about the electronic structure of the molecules. Figure 6.23a shows dI/dU spectra measured on a high topographic feature which does not emit light (dark molecule, green curve), and on Au(111) (black curve). The dark molecule shows a broad peak at a negative bias centered around -1.80 V. This broad peak originates from the molecule itself because on Au(111), no apparent feature is present. At positive bias, the dark molecule is featureless and shows low conductivity.

On a Tpdx-nDNI(A)/Tpdx-nDNI(B) molecule, dI/dU spectra were measured, and sharp peaks in the positive and negative bias voltages were observed, see Fig. 6.23b. For Tpdx-nDNI(A), the peak at the positive bias voltage is centered around 2.62 V whereas for Tpdx-nDNI(B), this peak is centered at a lower bias of 2.42 V. At negative bias, Tpdx-nDNI(A) has a broad peak feature at -2.69 V and if measured in larger bias range, this peak would be more apparent. However, at higher biases, molecule was not stable. Tpdx-nDNI(B) shows a clear peak centered around -2.35 V. These peaks at positive/negative bias can be assigned to the frontier orbitals of the molecules (LUMO/HOMO). The

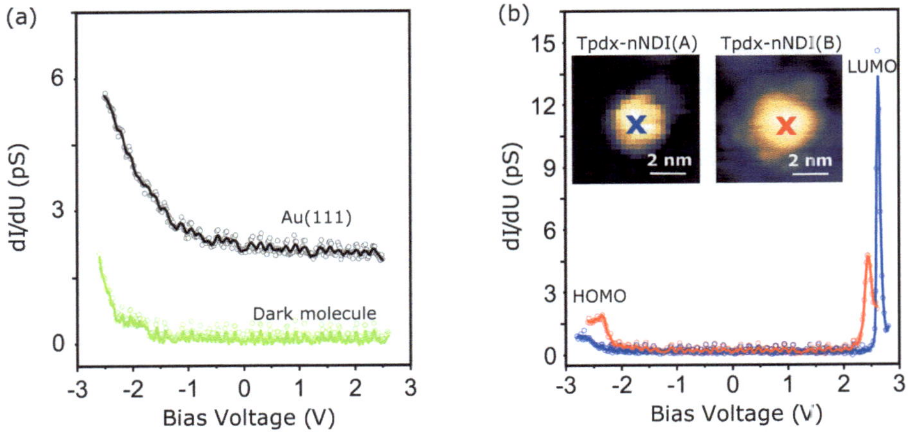

Figure 6.23: Differential conductance measurements. (a) dI/dU spectra measured on Au(111) (black line) (stabilization at -2.60 V), and on a dark molecule (green line). **(b)** dI/dU spectra on a Tpdx-nNDI(A) (blue line), and on a Tpdx-nNDI(B) (red line). Insets show the topography of the molecules ($U = 1.5$ V, $I = 2$ pA). The tip position is marked with blue/red cross. Lock-in parameters are: stabilization at $U = -2.60$ V, $I = 50$ pA and $U = -2.80$ V, $I = 30$ pA for two measurements, respectively. Lock-in modulation for all measurements is 20 mV.

dI/dU spectra suggest that the light emitting Tpdx-nNDI is decoupled from the Au(111) surface and the molecules with different emission types show different orders of decoupling which could also be associated with their difference in emission quantum efficiencies.

6.3 Characteristics of Tpdx-nNDI and Tpdx-sNDI Emissions Lines

It was mentioned in the section 6.2.2 that the peak energies of the emissions from Tpdx-sNDI and Tpdx-nNDI are sensitive on the measurement positions as well as on the applied bias voltage. In the following two sections, the spatial variation of Q_1 peak (Tpdx-nNDI(A)) and the effect of bias voltage on high energy peaks of Tpdx-sNDI and Tpdx-nNDI will be discussed.

6.3.1 Spatial Distribution

Figure 6.24 shows a spectrum recorded on an isolated Tpdx-nNDI(A) molecule. The observed quantum efficiency is 1.1×10^{-3} photons/electron. This is the highest quan-

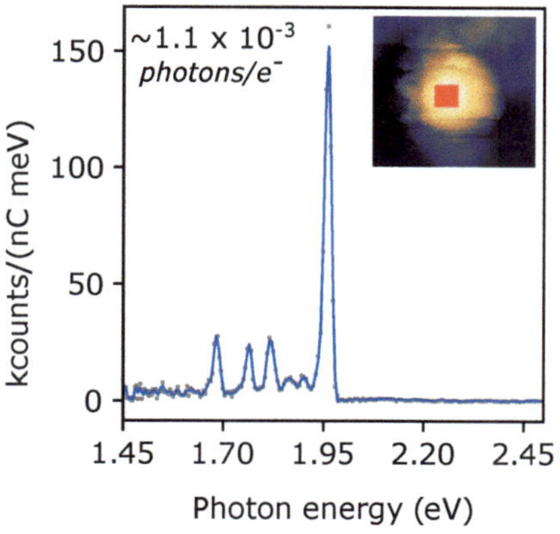

Figure 6.24: Highest quantum efficiency. Spectrum recorded on an Tpdx-nNDI ($U = 2.16\,\text{V}$, $I = 5\,\text{pA}$, $t = 0.5\,\text{s}$). Inset shows the topography of the molecule ($U = 1.8\,\text{V}$, $I = 2\,\text{pA}$).

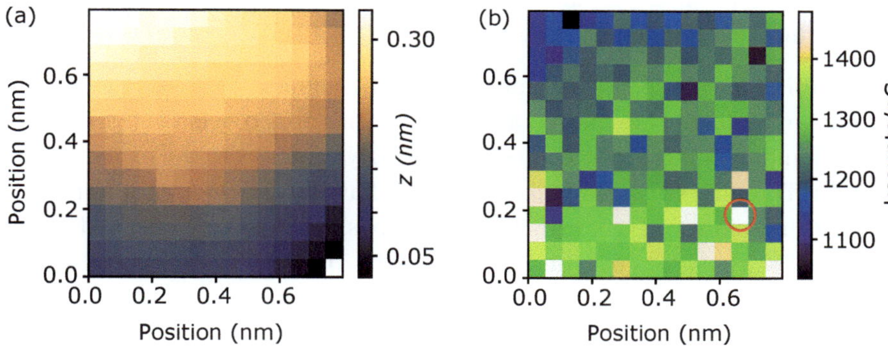

Figure 6.25: Spatial variation of Q peak intensities. (a) Topography of the part of the molecule shown in the inset of Fig. 6.24 ($U = 2.16\,\text{V}$, $I = 5\,\text{pA}$). **(b)** Photon map recorded on the region shown in **a** ($U = 2.16\,\text{V}$, $I = 5\,\text{pA}$, $t = 0.5\,\text{s}$). Red circle shows the pixel with highest observed quantum efficiency.

tum efficiency recorded for such self-decoupled molecules. This spectrum is a part of so called photon map shown in Fig. 6.25b. This photon map shows the integrated intensity of the spectrum in the energy range of $1.94\,\text{eV}$ - $1.97\,\text{eV}$ plotted over a grid of 15×15 pixels. From the photon map, it is observed that intensity of the Q_1 peak varies over the molecule (topography in Fig. 6.25b) and a gradient is observed in the $y-$axis direction. The topography shown in Fig. 6.25a is small part of the molecule (marked with a red rectangle in the inset topography of Fig. 6.24). It shows that the Q peak has a higher emission towards the edge of the molecule. This observation hints that the molecular transition dipole is not perfectly perpendicular to the substrate and out coupling is more efficient on the edge of the molecule.

Table 6.1: High energy peaks of Tpdx-sNDI and Tpdx-nNDI

Molecule	(A)	(B)
Tpdx-sNDI	2.235/Q	2.252/Q^+
Tpdx-nNDI	2.231/Q_2	2.253/Q^+

6.3.2 Variation in Peak Positions with Applied Field

The STML measurements on Tpdx-sNDI and Tpdx-nNDI reveal that they both show two different kinds of emissions. Interestingly, both types of molecules have a high energy peak, which is centered in the energy window of 2.224 eV - 2.255 eV (see table 6.1). Just from the values in the table, it is not clear that if there is any correlation between different emissions. Moreover, since they are from different molecules, any correlation would be contrary to the expectations.

To confirm this, in Fig. 6.26, the peak positions of these four high energy emission lines (mentioned in table 6.1) are plotted as a function of applied bias voltage for spectra recorded on many different molecules. Surprisingly, contrary to the expectation, the emissions of Tpdx-nDNI(A) and Tpdx-sDNI(A) behave similarly with the applied bias, whereas the emissions of Tpdx-nDNI(B) and Tpdx-sDNI(B) behave similarly with the applied bias. Peak positions of Q and Q_2 lines of the Tpdx-nDNI(A) and Tpdx-sDNI(A) show a blue shift with the increasing bias voltage. The peak positions of Q^+ lines of the Tpdx-nDNI(B) and Tpdx-sDNI(B) molecules do not show any change with increasing bias voltage.

This is a complex behaviour to interpret, but, one fact that could help to understand is: in both molecules (Tpdx-sNDI and Tpdx-nNDI) the core functional group is the same, which is the -NDI. If these high energy peaks are transitions related to the NDI core, then they can appear in the emissions of Tpdx-nDNI(B) and Tpdx-sDNI(B). This would suggest that the molecules go under some kind of chemical modification due to the presence of the tip and the substituted side groups are no longer affecting the emission of the molecules.

This chemical change via the tip is very subtle and not apparent in the STM measurements. However, the optical fingerprint does support the argument of chemical modification of the molecules via the STM tip.

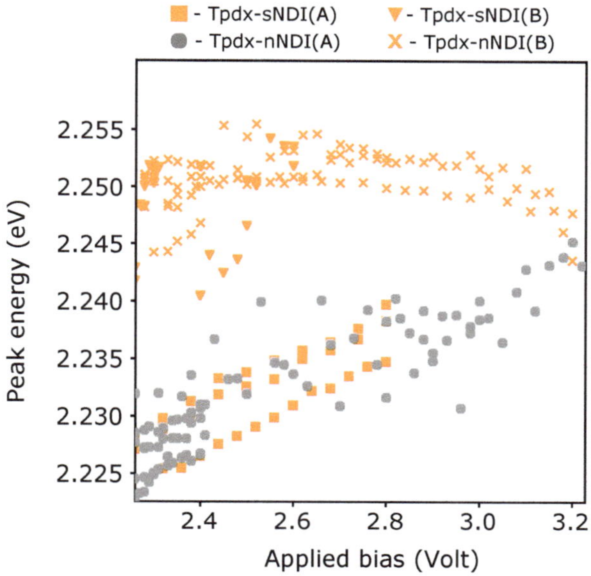

Figure 6.26: Variations in peak positions with applied bias. Variations in the high energy peaks of Tpdx-sNDI and Tpdx-nNDI for their two types of emissions with applied sample bias.

6.4 Summary and Discussion

In this chapter, light emission from self-decoupled molecules has been studied. The necessary electronic decoupling of light emitting chromophores (sNDI and nNDI) from the underneath metal surface was achieved by anchoring them to a new type of tripodal platform (Tpdx).

The chapter starts with the STM and STML studies of Tpdx-sNDI molecules directly deposited on Au(111) surface. Molecules showed two different kinds of emissions. Based on the emission profile, Tpdx-sNDI molecules can be categorized as Tpdx-sNDI(A) and Tpdx-sNDI(B) molecules. Bias dependent measurements showed that the molecules are excited by IET via plasmons and excitation by electron-hole pair injection is unlikely. Furthermore, differential conductance measurements showed the clear differences between the electronic structure of molecules which emit light and the molecules which did not (dark molecules). Also, the spatial dependence of optical emission spectra showed no sign of plasmon-exciton coupling when the STM tip was in close proximity of the molecule. This part of the chapter establishes the fact that the introduced tripodal platform successfully decouples the light-emitting chromophore.

In order to further test the reproducibility and versatility of the tripodal platform, another molecule, Tpdx-nNDI has been studied. Some of the measurements on these molecules were done with higher spectral resolution, hence very sharp emission lines were observed (linewidth of $\sim 7\,\mathrm{meV}$). This molecule also showed two kinds of emissions. For this molecule, both types of emissions have bipolar nature with applied bias. Quantum efficiencies are again different for both emissions with Tpdx-nNDI(A) showing record high emission efficiency of 1.1×10^{-3} photons/electron for such self-decoupled molecules.

From the bias and current-dependent measurements on Tpdx-nNDI(A), it was observed that two different excitation processes are involved in two different bias voltage regimes. If the energy of the exciting electron was less than the energy of the emitted photon, $eU < h\nu$, then the UCEL was observed. Once the situation was $eU = h\nu$, then the normal emissions from the molecules were observed (excited by IET). The optical emission spectrum of the molecule showed the vibrational fingerprint of the molecule. Not only this, the high resolution spectrum of the molecule shows the phenomenon called hot electroluminescence, where we observed emissions from several radiative transitions directly taking place from the excited vibrational states of an excited electronic state to the vibrational ground state of the electronic ground state, hence violating the Kasha's rule.

Contrary to the previous reports on light emission from self-decoupled molecules, it is really remarkable that the Tpdx-nNDI molecule not only shows molecular emission but also shows many intriguing phenomena such as vibrationally resolved fluorescence, hot electroluminescence, UCEL, which were commonly observed in stable systems of planer molecules like H_2Pc, ZnPc, MnPc etc. Furthermore, we have also observed phenomena like large Stark shifts (up to even $200\,\mathrm{meV}$), and switching between different emission types.

Even though the employed tripodal platform successfully decoupled the light emitting chromophores, it is still not possible to excite the molecules by electron-hole pair injection to achieve electroluminescence, hence achieving single molecule OLEDs. To observe electroluminescence from NDIs, a high bias voltage is required ($> 3.5\,\mathrm{V}$ for Tpdx-nNDI), which the molecules could not survive. Therefore, these findings provide a clear guideline for the future work, that is, to design chromophores which allow electroluminescence at lower biases.

7 Conclusion and Outlook

The aim of this thesis was to study the optoelectronic properties of single molecules in an STM junction with the purpose to understand the mechanism behind efficient emission. The findings of this thesis provide pathways for efficient light emission which are not specific to a molecule and can be extended to any chromophore.

The first result chapter (4) discusses how the emission efficiency of a H_2Pc molecule can be boosted by charging. The STML results indicated that the emission efficiencies of the charged molecules is ≈ 19 times more than the neutral. The quantum chemical calculations show that this boost was observed due to the development of a vertical transfer dipole moment in the presence of the STM tip. This indicates that the efficiency can be enhanced by orders of magnitude as it depends on the magnitude of the vertical component of the dipole moment. In this experiment, fluorescence quenching of the molecule was prohibited by inserting a thin insulating layer of NaCl.

From the application point of view, the salt layer needs to be replaced by the molecule itself. Moreover, molecules should also offer a large vertical dipole component for efficient emission. In the next results chapter (5), the idea of functionalizing molecules for self-decoupling was implemented. Three chromophores -hNDI, -sNDI and -nNDI anchored to a functional unit, which works as the tripodal platform (Tpd) to decouple the chromophores were tried. These molecules could not be sublimed because they do not survive high temperatures. To overcome this issue, a spray deposition technique was used, which was developed in our group. The sprayed sample was then annealed to promote the chemical reaction with the substrate. We observed that the adsorptions geometries of the molecules strongly depend on the annealing temperature. At lower temperatures, molecules formed clusters. These clusters showed molecular luminescence spectra with an internal quantum yield of 6×10^{-3} photons/electron. Further annealing the sample resulted in the formation of well-ordered molecular islands. In this case, molecules were hybridized with the substrate and did not show molecular luminescence.

In the ordered islands of Tpd-sNDI, STM tip was used to manipulate individual molecules and hence modifying the electronic decoupling of the molecules. Surprisingly, a small change in the adsorption geometry resulted in spectra with sharp emission lines. With the help of calculations, these emission lines were assigned to the excitation of a singly charged cationic dimer of Tpd-sNDI. The presented study pioneers the well-defined manipulation of the luminescence quantum yield of individual molecules directly adsorbed on a metal surface. These results prove the principle of manipulating the fluorescence response of individual molecules in a matrix by controlling the degree of decoupling with the underlying substrate.

Motivated by the results on the switched Tpd-sNDI, a new anchored group (Tpdx) was functionalized with the -sNDI chromophore. The STML measurements showed vibrationally resolved molecular luminescence. The main transition line matches the photoluminescence spectra recorded in solution. Differential conductance measurements showed that the light emitting molecules were electronically decoupled from the substrate. The versatility of the anchor group allowed us to functionalize different chromophores. In light of this, another chromophore -nNDI was functionalized to Tpdx. The optical emission spectra recorded on this molecule showed a high emission efficiency of 1.1×10^{-3} photons/electron. The stability of the molecules allowed us to perform spatially resolved fluorescence spectroscopy. Furthermore, current-dependent measurements on the Tpdx-nNDI allowed us to directly see the effect of electrode on the fluorescence of the molecule. Initially, with the increasing current, the intensity of the emission lines increased. At sufficiently high currents ($\approx$71 pA), the tip is close to the molecule and the fluorescence is quenched. Tpdx-nNDI also showed overbias emission, where emitted photons have higher energies than the exciting electrons. The emission spectra showed the phenomenon called hot electroluminescence, where due to the high density of electromagnetic modes in the cavity, molecules de-excite directly from the excited vibrational state of the excited electronic state to the ground electronic state. This allowed us to get the spectroscopic fingerprint of the excited states.

We also observed that with the help of the STM tip, it was possible to chemically modify the molecules. This was reflected in the corresponding emission spectrum. In the presence of the STM tip, both molecules, Tpdx-sNDI and Tpdx-nNDI switch to a more stable state and produce identical spectra. This hints that the molecules are modified in such a way that they end up in similar state. Structural change was very small and it was not apparent in the topography.

The findings of this thesis provide guidelines for future experiments to enhance the performance of single molecule light emitting device. The outlook can be summarized in following points:

- Essentially, molecule has to provide a large vertical dipole component in the junction to effectively couple to the electromagnetic field.
- Emission spectra recorded on the clusters of Tpd-hNDI, Tpd-sNDI and Tpd-nNDI showcase their potential for highly efficient OLED devices.

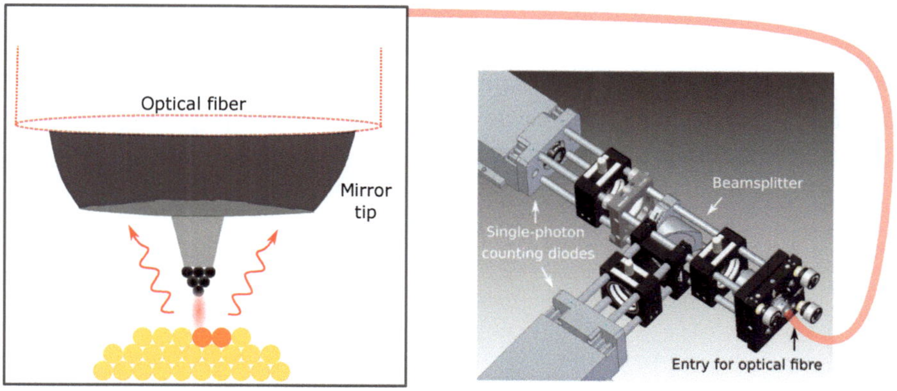

Figure 7.1: Upgraded STML setup for TR measurements. Left STML setup with mirror tip. Collected light is out-couple by an optical fiber. **Right** Hanbury Brown–Twiss interferometer: The Light exiting from the fiber is collimated via a lense. A beamsplitter is then used to send equal intesity into the two single-photon avalanche photodiodes (SPADs) from company *Excelitas*.

- The switching behaviour of the Tpd-sNDI shows the possibility of combining the self-assembly of the molecular layer and the possibility to manipulate individual molecules with voltage pulses in order to emit light into the far-field. This gives a conceptual size limit of a display pixel. In the future, different functional groups can be used to gain more control over the switching mechanism.
- Experiments of the Tpdx-sNDI and Tpdx-nNDI suggest that the excitation of molecules via IET mechanism, which means this not a OLED device in true sense. To achieve a single molecule OLED, another type of chromophore with a small HOMO-LUMO gap can be functionalized with the tripodal platform.
- The Tpdx platform can be used for various other chromophores, making it a perfect candidate to achieve a chemically tunable photon source.

Further insight can be gained into the the mechanism behind electroluminescence by time-resolved measurements. Recently, a Hanbury Brown-Twiss interferometer was installed with the γ-STM to perform time-correlated single photon counting (see Fig. 7.1). The setup was built and tested by Lukas Gerhard, Gabriel Derenbach and Kevin Edelmann. Contrary to the STM measurements, time-resolved (TR) measurements provide information on much shorter time scale. This helps us to study phenomena which are not apparent in the tunnelling current but are apparent in the emitted light. This kind of setup has been used to study the fluctuations in the tunnelling junction [169–171], single photon behaviour in a molecular crystal [172], and single molecules decoupled via NaCl [20]. These results clearly show that the time-resolved measurements on single self-decoupled molecules like Tpdx-sNDI and Tpdx-nNDI can give the information about dynamics at atomic scale [173]. Thus making them promising candidates for producing single photons with narrow line emission, a clear characteristic for quantum computation applications. Future experiments will deal with the TR-STML measurements on a variety of chromophores mounted on the well-established Tpdx platform to achieve single photon molecular sources.

List of Figures

Bibliography

[1] W. E. Moerner. “Single-molecule spectroscopy, imaging, and photocontrol: foundations for super-resolution microscopy (nobel lecture)”. In: *Angewandte Chemie International Edition* 54.28 (2015), pp. 8067–8093.

[2] W. E. Moerner and L. Kador. “Optical detection and spectroscopy of single molecules in a solid”. In: *Physical Review Letters* 62.21 (1989), p. 2535.

[3] K. Kuhnke, C. Grosse, P. Merino, and K. Kern. “Atomic-scale imaging and spectroscopy of electroluminescence at molecular interfaces”. In: *Chemical Reviews* 117.7 (2017), pp. 5174–5222.

[4] E. Abbe. “Beiträge zur Theorie des Mikroskops und der mikroskopischen Wahrnehmung”. In: *Archiv für Mikroskopische Anatomie* 9.1 (1873), pp. 413–468.

[5] X. Qiu, G. Nazin, and W. Ho. “Vibrationally resolved fluorescence excited with submolecular precision”. In: *Science* 299.5606 (2003), pp. 542–546.

[6] Z. Dong, X. Zhang, H. Gao, Y. Luo, C. Zhang, L. Chen, R. Zhang, X. Tao, Y. Zhang, J. Yang, et al. “Generation of molecular hot electroluminescence by resonant nanocavity plasmons”. In: *Nature Photonics* 4.1 (2010), pp. 50–54.

[7] Y. Zhang, Y. Luo, Y. Zhang, Y.-J. Yu, Y.-M. Kuang, L. Zhang, Q.-S. Meng, Y. Luo, J.-L. Yang, Z.-C. Dong, et al. “Visualizing coherent intermolecular dipole–dipole coupling in real space”. In: *Nature* 531.7596 (2016), p. 623.

[8] G. Binnig, H. Rohrer, C. Gerber, and E. Weibel. “Tunneling through a controllable vacuum gap”. In: *Applied Physics Letters* 40.2 (1982), pp. 178–180.

[9] G. Binnig, H. Rohrer, C. Gerber, and E. Weibel. “Surface studies by scanning tunneling microscopy”. In: *Physical Review Letters* 49.1 (1982), p. 57.

[10] J. Gimzewski, B. Reihl, J. Coombs, and R. Schlittler. “Photon emission with the scanning tunneling microscope”. In: *Zeitschrift für Physik B Condensed Matter* 72.4 (1988), pp. 497–501.

[11] R. Berndt, J. K. Gimzewski, and P. Johansson. "Inelastic tunneling excitation of tip-induced plasmon modes on noble-metal surfaces". In: *Physical Review Letters* 67.27 (1991), p. 3796.

[12] R. Berndt and J. Gimzewski. "Injection luminescence from CdS (11-20) studied with scanning tunneling microscopy". In: *Physical Review B* 45.24 (1992), p. 14095.

[13] R. Berndt, R. Gaisch, J. Gimzewski, B. Reihl, R. Schlittler, W. Schneider, and M. Tschudy. "Photon emission at molecular resolution induced by a scanning tunneling microscope". In: *Science* 262.5138 (1993), pp. 1425–1427.

[14] N. Schneider, F. Matino, G. Schull, S. Gabutti, M. Mayor, and R. Berndt. "Light emission from a double-decker molecule on a metal surface". In: *Physical Review B* 84.15 (2011), p. 153403.

[15] S.-E. Zhu, Y.-M. Kuang, F. Geng, J.-Z. Zhu, C.-Z. Wang, Y.-J. Yu, Y. Luo, Y. Xiao, K.-Q. Liu, Q.-S. Meng, et al. "Self-decoupled porphyrin with a tripodal anchor for molecular-scale electroluminescence". In: *Journal of the American Chemical Society* 135.42 (2013), pp. 15794–15800.

[16] M. C. Chong, G. Reecht, H. Bulou, A. Boeglin, F. Scheurer, F. Mathevet, and G. Schull. "Narrow-line single-molecule transducer between electronic circuits and surface plasmons". In: *Physical Review Letters* 116.3 (2016), p. 036802.

[17] T. Ijaz, B. Yang, R. Wang, J. Zhu, A. Farrukh, G. Chen, G. Franc, Y. Zhang, A. Gourdon, and Z. Dong. "Self-decoupled tetrapodal perylene molecules for luminescence studies of isolated emitters on Au (111)". In: *Applied Physics Letters* 115.17 (2019), p. 173101.

[18] H. Imada, K. Miwa, M. Imai-Imada, S. Kawahara, K. Kimura, and Y. Kim. "Real-space investigation of energy transfer in heterogeneous molecular dimers". In: *Nature* 538.7625 (2016), p. 364.

[19] Y. Zhang, Q.-S. Meng, L. Zhang, Y. Luo, Y.-J. Yu, B. Yang, Y. Zhang, R. Esteban, J. Aizpurua, Y. Luo, et al. "Sub-nanometre control of the coherent interaction between a single molecule and a plasmonic nanocavity". In: *Nature Communications* 8 (2017), p. 15225.

[20] L. Zhang, Y.-J. Yu, L.-G. Chen, Y. Luo, B. Yang, F.-F. Kong, G. Chen, Y. Zhang, Q. Zhang, Y. Luo, et al. "Electrically driven single-photon emission from an isolated single molecule". In: *Nature Communications* 8.1 (2017), pp. 1–7.

[21] H. Imada, K. Miwa, M. Imai-Imada, S. Kawahara, K. Kimura, and Y. Kim. "Single-molecule investigation of energy dynamics in a coupled plasmon-exciton system". In: *Physical Review Letters* 119.1 (2017), p. 013901.

[22] B. Doppagne, M. C. Chong, E. Lorchat, S. Berciaud, M. Romeo, H. Bulou, A. Boeglin, F. Scheurer, and G. Schull. "Vibronic spectroscopy with submolecular resolution from STM-induced electroluminescence". In: *Physical Review Letters* 118.12 (2017), p. 127401.

[23] B. Doppagne, M. C. Chong, H. Bulou, A. Boeglin, F. Scheurer, and G. Schull. "Electrofluorochromism at the single-molecule level". In: *Science* 361.6399 (2018), pp. 251–255.

[24] G. Chen, Y. Luo, H. Gao, J. Jiang, Y. Yu, L. Zhang, Y. Zhang, X. Li, Z. Zhang, and Z. Dong. "Spin-triplet-mediated up-conversion and crossover behavior in single-molecule electroluminescence". In: *Physical Review Letters* 122.17 (2019), p. 177401.

[25] K. Miwa, H. Imada, M. Imai-Imada, K. Kimura, M. Galperin, and Y. Kim. "Many-body state description of single-molecule electroluminescence driven by a scanning tunneling microscope". In: *Nano Letters* 19.5 (2019), pp. 2803–2811.

[26] K. Kaiser, L. Gross, and F. Schulz. "A single-molecule chemical reaction studied by high-resolution atomic force microscopy and scanning tunneling microscopy induced light emission". In: *ACS Nano* 13.6 (2019), pp. 6947–6954.

[27] J. Dolezal, P. Merino, J. Redondo, L. Ondic, A. Cahlik, and M. Svec. "Charge carrier injection electroluminescence with CO-functionalized tips on single molecular emitters". In: *Nano Letters* 19.12 (2019), pp. 8605–8611.

[28] B. Doppagne, T. Neuman, R. Soria-Martinez, L. E. P. López, H. Bulou, M. Romeo, S. Berciaud, F. Scheurer, J. Aizpurua, and G. Schull. "Single-molecule tautomerization tracking through space-and time-resolved fluorescence spectroscopy". In: *Nature Nanotechnology* 15.3 (2020), pp. 207–211.

[29] V. Rai, L. Gerhard, Q. Sun, C. Holzer, T. Repän, M. Krstić, L. Yang, M. Wegener, C. Rockstuhl, and W. Wulfhekel. "Boosting light emission from single hydrogen phthalocyanine molecules by charging". In: *Nano Letters* 20.10 (2020), pp. 7600–7605.

[30] T.-C. Hung, B. Kiraly, J. H. Strik, A. A. Khajetoorians, and D. Wegner. "Plasmon-driven motion of an individual molecule". In: *Nano Letters* (2021).

[31] N. Balzer, J. Lukášek, M. Valášek, V. Rai, Q. Sun, L. Gerhard, W. Wulfhekel, and M. Mayor. "Synthesis and surface behavior of NDI-chromophores mounted on

a tripodal scaffold: towards self-Decoupled chromophores for single molecule electroluminescence". In: *Chemistry–A European Journal* (2021).

[32] D. J. Griffiths and D. F. Schroeter. *Introduction to quantum mechanics*. Cambridge University Press, 2018.

[33] P. K. Hansma and J. Tersoff. "Scanning tunneling microscopy". In: *Journal of Applied Physics* 61.2 (1987), R1–R24.

[34] G. Binnig and D. P. Smith. "Single-tube three-dimensional scanner for scanning tunneling microscopy". In: *Review of Scientific Instruments* 57.8 (1986), pp. 1688–1689.

[35] J. Chen. *Introduction to Scanning Tunneling Microscopy Third Edition*. Vol. 69. Oxford University Press, USA, 2021.

[36] C. Joachim, J. K. Gimzewski, R. R. Schlittler, and C. Chavy. "Electronic transparence of a single C 60 molecule". In: *Physical Review Letters* 74.11 (1995), p. 2102.

[37] E. Scheer, N. Agraït, J. C. Cuevas, A. L. Yeyati, B. Ludoph, A. Martín-Rodero, G. R. Bollinger, J. M. van Ruitenbeek, and C. Urbina. "The signature of chemical valence in the electrical conduction through a single-atom contact". In: *Nature* 394.6689 (1998), pp. 154–157.

[38] J. Fisher and I. Giaever. "Tunneling through thin insulating layers". In: *Journal of Applied Physics* 32.2 (1961), pp. 172–177.

[39] J. Bardeen. "Tunnelling from a many-particle point of view". In: *Physical Review Letters* 6.2 (1961), p. 57.

[40] J. Tersoff and D. Hamann. "Theory and application for the scanning tunneling microscope". In: *Physical Review Letters* 50.25 (1983), p. 1998.

[41] C. J. Chen. "Origin of atomic resolution on metal surfaces in scanning tunneling microscopy". In: *Physical Review Letters* 65.4 (1990), p. 448.

[42] S. Lounis. "Theory of scanning tunneling microscopy". In: *arXiv preprint arXiv:1404.0961* (2014).

[43] J. Lambe and R. Jaklevic. "Molecular vibration spectra by inelastic electron tunneling". In: *Physical Review* 165.3 (1968), p. 821.

[44] S. Khanna and J. Lambe. "Inelastic electron tunneling spectroscopy". In: *Science* 220.4604 (1983), pp. 1345–1351.

[45] W. Ho. "Single-molecule chemistry". In: *The Journal of Chemical Physics* 117.24 (2002), pp. 11033–11061.

[46] B. Stipe, M. Rezaei, and W. Ho. "Inducing and viewing the rotational motion of a single molecule". In: *Science* 279.5358 (1998), pp. 1907–1909.

[47] J. Lambe and S. McCarthy. "Light emission from inelastic electron tunneling". In: *Physical Review Letters* 37.14 (1976), p. 923.

[48] A. Heinrich, J. Gupta, C. Lutz, and D. Eigler. "Single-atom spin-flip spectroscopy". In: *Science* 306.5695 (2004), pp. 466–469.

[49] K. Meguro, K. Sakamoto, R. Arafune, M. Satoh, and S. Ushioda. "Origin of multiple peaks in the light emission spectra of a Au (111) surface induced by the scanning tunneling microscope". In: *Physical Review B* 65.16 (2002), p. 165405.

[50] C. Kittel. "Introduction to solid state physics". In: (1976).

[51] A. V. Zayats, I. I. Smolyaninov, and A. A. Maradudin. "Nano-optics of surface plasmon polaritons". In: *Physics Reports* 408.3-4 (2005), pp. 131–314.

[52] W. L. Barnes, A. Dereux, and T. W. Ebbesen. "Surface plasmon subwavelength optics". In: *Nature* 424.6950 (2003), pp. 824–830.

[53] C. L. Smith, N. Stenger, A. Kristensen, N. A. Mortensen, and S. I. Bozhevolnyi. "Gap and channeled plasmons in tapered grooves: a review". In: *Nanoscale* 7.21 (2015), pp. 9355–9386.

[54] F. Rossel, M. Pivetta, and W.-D. Schneider. "Luminescence experiments on supported molecules with the scanning tunneling microscope". In: *Surface Science Reports* 65.5 (2010), pp. 129–144.

[55] R. Rendell and D. Scalapino. "Surface plasmons confined by microstructures on tunnel junctions". In: *Physical Review B* 24.6 (1981), p. 3276.

[56] J. Aizpurua, G. Hoffmann, S. Apell, and R. Berndt. "Electromagnetic coupling on an atomic scale". In: *Physical Review Letters* 89.15 (2002), p. 156803.

[57] I. Romero, J. Aizpurua, G. W. Bryant, and F. J. G. De Abajo. "Plasmons in nearly touching metallic nanoparticles: singular response in the limit of touching dimers". In: *Optics Express* 14.21 (2006), pp. 9988–9999.

[58] G. Schull, N. Néel, P. Johansson, and R. Berndt. "Electron-plasmon and electron-electron interactions at a single atom contact". In: *Physical Review Letters* 102.5 (2009), p. 057401.

[59] N. L. Schneider, G. Schull, and R. Berndt. "Optical probe of quantum shot-noise reduction at a single-atom contact". In: *Physical Review Letters* 105.2 (2010), p. 026601.

[60] L. Douillard and F. Charra. "High-resolution mapping of plasmonic modes: photoemission and scanning tunnelling luminescence microscopies". In: *Journal of Physics D: Applied Physics* 44.46 (2011), p. 464002.

[61] E. M. Purcell, H. C. Torrey, and R. V. Pound. "Resonance absorption by nuclear magnetic moments in a solid". In: *Physical Review* 69.1-2 (1946), p. 37.

[62] J. Doležal, S. Canola, P. Merino, and M. Švec. "Exciton-Trion Conversion Dynamics in a Single Molecule". In: *ACS Nano* 15.4 (2021), pp. 7694–7699.

[63] Z.-C. Dong, X.-L. Guo, A. Trifonov, P. Dorozhkin, K. Miki, K. Kimura, S. Yokoyama, and S. Mashiko. "Vibrationally resolved fluorescence from organic molecules near metal surfaces in a scanning tunneling microscope". In: *Physical Review Letters* 92.8 (2004), p. 086801.

[64] K. Miwa, A. M. Najarian, R. L. McCreery, and M. Galperin. "Hubbard Nonequilibrium Green's Function Analysis of Photocurrent in Nitroazobenzene Molecular Junction". In: *The Journal of Physical Chemistry Letters* 10.7 (2019), pp. 1550–1557.

[65] N. L. Schneider and R. Berndt. "Plasmonic excitation of light emission and absorption by porphyrine molecules in a scanning tunneling microscope". In: *Physical Review B* 86.3 (2012), p. 035445.

[66] K. Sakamoto, K. Meguro, R. Arafune, M. Satoh, Y. Uehara, and S. Ushioda. "Light emission spectra of the monolayer-island of C60 molecules on Au (111) induced by scanning tunneling microscope". In: *Surface Science* 502 (2002), pp. 149–155.

[67] W. Deng, D. Fujita, T. Ohgi, S. Yokoyama, K. Kamikado, and S. Mashiko. "STM-induced photon emission from self-assembled porphyrin molecules on a Cu (100) surface". In: *The Journal of Chemical Physics* 117.10 (2002), pp. 4995–5000.

[68] Z.-C. Dong, A. Trifonov, X.-L. Guo, K. Amemiya, S. Yokoyama, T. Kamikado, T. Yamada, S. Mashiko, and T. Okamoto. "Tunneling electron induced photon emission from monolayered H2TBP porphyrin molecules on Cu (100)". In: *Surface Science* 532 (2003), pp. 237–243.

[69] X.-L. Guo, Z.-C. Dong, A. Trifonov, S. Yokoyama, S. Mashiko, and T. Okamoto. "Light emission from organic molecules on metal substrates induced by tunneling currents". In: *Japanese Journal of Applied Physics* 42.11R (2003), p. 6937.

[70] G. Hoffmann, T. Maroutian, and R. Berndt. "Color view of atomic highs and lows in tunneling induced light emission". In: *Physical Review Letters* 93.7 (2004), p. 076102.

[71] F. Geng, Y. Zhang, Y. Yu, Y. Kuang, Y. Liao, Z. Dong, and J. Hou. "Modulation of nanocavity plasmonic emission by local molecular states of c 60 on au (111)". In: *Optics Express* 20.24 (2012), pp. 26725–26735.

[72] N. Balzer. *Design, Synthesis and Evolution of Functional Tripodal Platforms for Single Molecule Devices.* Vol. 14258. University of Basel, Faculty of Science, 2021.

[73] K. Edelmann, L. Gerhard, M. Winkler, L. Wilmes, V. Rai, M. Schumann, C. Kern, M. Meyer, M. Wegener, and W. Wulfhekel. "Light collection from a low-temperature scanning tunneling microscope using integrated mirror tips fabricated by direct laser writing". In: *Review of Scientific Instruments* 89.12 (2018), p. 123107.

[74] K. Edelmann. *Electroluminescence from Plasmonic Excitations in a Scanning Tunnelling Microscope.* Vol. 24. KIT Scientific Publishing, 2019.

[75] S. Pan, E. W. Hudson, and J. Davis. "3 He refrigerator based very low temperature scanning tunneling microscope". In: *Review of Scientific Instruments* 70.2 (1999), pp. 1459–1463.

[76] L. Zhang, T. Miyamachi, T. Tomanić, R. Dehm, and W. Wulfhekel. "A compact sub-Kelvin ultrahigh vacuum scanning tunneling microscope with high energy resolution and high stability". In: *Review of Scientific Instruments* 82.10 (2011), p. 103702.

[77] C. Renner, P. Niedermann, A. Kent, and O. Fischer. "A vertical piezoelectric inertial slider". In: *Review of Scientific Instruments* 61.3 (1990), pp. 965–967.

[78] J. Gimzewski, J. Sass, R. Schlitter, and J. Schott. "Enhanced photon emission in scanning tunnelling microscopy". In: *EPL (Europhysics Letters)* 8.5 (1989), p. 435.

[79] D. Abraham, A. Veider, C. Schönenberger, H. Meier, D. Arent, and S. Alvarado. "Nanometer resolution in luminescence microscopy of III-V heterostructures". In: *Applied Physics Letters* 56.16 (1990), pp. 1564–1566.

[80] G. Hoffmann, J. Kröger, and R. Berndt. "Color imaging with a low temperature scanning tunneling microscope". In: *Review of Scientific Instruments.* 73.2 (2002), pp. 305–309.

[81] J. Lee, S. M. Perdue, A. Rodriguez Perez, and V. A. Apkarian. "Vibronic motion with joint angstrom–femtosecond resolution observed through Fano progressions recorded within one molecule". In: *ACS Nano* 8.1 (2014), pp. 54–63.

[82] K. Kuhnke, A. Kabakchiev, W. Stiepany, F. Zinser, R. Vogelgesang, and K. Kern. "Versatile optical access to the tunnel gap in a low-temperature scanning tunneling microscope". In: *Review of Scientific Instruments.* 81.11 (2010), p. 113102.

[83] L. Chen, C. Zhang, R. Zhang, X. Zhang, and Z. Dong. "Note: optical optimization for ultrasensitive photon mapping with submolecular resolution by scanning tunneling microscope induced luminescence". In: *Review of Scientific Instruments* 84.6 (2013), p. 066106.

[84] J. G. Keizer, J. K. Garleff, and P. M. Koenraad. "Simple and efficient scanning tunneling luminescence detection at low-temperature". In: *Review of Scientific Instruments.* 80.12 (2009), p. 123704.

[85] S. Ushioda. "STM light emission spectroscopy of Si surfaces". In: *Solid State Communications* 84.1-2 (1992), pp. 173–176.

[86] N. Nilius, N. Ernst, and H.-J. Freund. "Photon emission spectroscopy of individual oxide-supported silver clusters in a scanning tunneling microscope". In: *Physical Review Letters* 84.17 (2000), p. 3994.

[87] Y. Khang, Y. Park, M. Salmeron, and E. R. Weber. "Low temperature ultrahigh vacuum cross-sectional scanning tunneling microscope for luminescence measurements". In: *Review of Scientific Instruments.* 70.12 (1999), pp. 4595–4599.

[88] M. J. Romero, J. van de Lagemaat, I. Mora-Sero, G. Rumbles, and M. M. Al-Jassim. "Imaging of resonant quenching of surface plasmons by quantum dots". In: *Nano Letters* 6.12 (2006), pp. 2833–2837.

[89] Y. Suzuki, H. Minoda, and N. Yamamoto. "STM light emission from Ag/Si (111)". In: *Surface Science* 438.1-3 (1999), pp. 297–304.

[90] M. J. Gallagher, S. Howells, L. Yi, T. Chen, and D. Sarid. "Photon emission from gold surfaces in air using scanning tunneling microscopy". In: *Surface Science* 278.3 (1992), pp. 270–280.

[91] C. Thirstrup, M. Sakurai, and M. Aono. "Photon emission STM using optical fiber bunches". In: *Journal of Surface Analysis* 4 (1988), pp. 152–158.

[92] R. Arafune, K. Sakamoto, K. Meguro, M. Satoh, A. Arai, and S. Ushioda. "Multiple-fiber collection system for scanning tunneling microscope light emission spectroscopy". In: *Japanese Journal of Applied Physics* 40.9R (2001), p. 5450.

[93] N. J. Watkins, J. P. Long, Z. H. Kafafi, and A. J. Mäkinen. "Fiber optic light collection system for scanning-tunneling-microscope-induced light emission". In: *Review of Scientific Instruments* 78.5 (2007), p. 053707.

[94] R. Berndt, R. R. Schlittler, and J. K. Gimzewski. "Photon emission scanning tunneling microscope". In: *Journal of Vacuum Science & Technology B: Microelectronics and Nanometer Structures Processing, Measurement, and Phenomena* 9.2 (1991), pp. 573–577.

[95] S. Kawata, H.-B. Sun, T. Tanaka, and K. Takada. "Finer features for functional microdevices". In: *Nature* 412.6848 (2001), pp. 697–698.

[96] M. Valášek, K. Edelmann, L. Gerhard, O. Fuhr, M. Lukas, and M. Mayor. "Synthesis of molecular tripods based on a rigid 9, 9′ -spirobifluorene scaffold". In: *The Journal of Organic Chemistry* 79.16 (2014), pp. 7342–7357.

[97] A. J. White, S. Tretiak, and M. Galperin. "Raman scattering in molecular junctions: a pseudoparticle formulation". In: *Nano Letters* 14.2 (2014), pp. 699–703.

[98] E. Ćavar, M.-C. Blüm, M. Pivetta, F. Patthey, M. Chergui, and W.-D. Schneider. "Fluorescence and phosphorescence from individual C 60 molecules excited by local electron tunneling". In: *Physical Review Letters* 95.19 (2005), p. 196102.

[99] J. Repp, G. Meyer, S. M. Stojković, A. Gourdon, and C. Joachim. "Molecules on insulating films: scanning-tunneling microscopy imaging of individual molecular orbitals". In: *Physical Review Letters* 94.2 (2005), p. 026803.

[100] A. Kabakchiev, K. Kuhnke, T. Lutz, and K. Kern. "Electroluminescence from individual pentacene nanocrystals". In: *ChemPhysChem* 11.16 (2010), pp. 3412–3416.

[101] J. Lee, S. M. Perdue, A. Rodriguez Perez, and V. A. Apkarian. "Vibronic motion with joint angstrom–femtosecond resolution observed through Fano progressions recorded within one molecule". In: *ACS Nano* 8.1 (2014), pp. 54–63.

[102] G. Reecht, F. Scheurer, V. Speisser, Y. J. Dappe, F. Mathevet, and G. Schull. "Electroluminescence of a polythiophene molecular wire suspended between a metallic surface and the tip of a scanning tunneling microscope". In: *Physical Review Letters* 112.4 (2014), p. 047403.

[103] K. Lauwaet, K. Schouteden, E. Janssens, C. Van Haesendonck, and P. Lievens. "Dependence of the NaCl/Au (111) interface state on the thickness of the NaCl layer". In: *Journal of Physics: Condensed Matter* 24.47 (2012), p. 475507.

[104] L. Gross. "Recent advances in submolecular resolution with scanning probe microscopy". In: *Nature Chemistry* 3.4 (2011), pp. 273–278.

[105] K. Miwa, H. Imada, S. Kawahara, and Y. Kim. "Effects of molecule-insulator interaction on geometric property of a single phthalocyanine molecule adsorbed on an ultrathin NaCl film". In: *Physical Review B* 93.16 (2016), p. 165419.

[106] M. Imai-Imada, H. Imada, K. Miwa, J. Jung, T. K. Shimizu, M. Kawai, and Y. Kim. "Energy-level alignment of a single molecule on ultrathin insulating film". In: *Physical Review B* 98.20 (2018), p. 201403.

[107] V. Bondybey and J. English. "Spectra of the h2 phthalocyanine in low-temperature matrixes". In: *Journal of the American Chemical Society* 101.13 (1979), pp. 3446–3450.

[108] C. Murray, N. Dozova, J. G. McCaffrey, N. Shafizadeh, W. Chin, M. Broquier, and C. Crépin. "Visible luminescence spectroscopy of free-base and zinc phthalocyanines isolated in cryogenic matrices". In: *Physical Chemistry Chemical Physics* 13.39 (2011), pp. 17543–17554.

[109] P. S. Fitch, C. A. Haynam, and D. H. Levy. "The fluorescence excitation spectrum of free base phthalocyanine cooled in a supersonic free jet". In: *The Journal of Chemical Physics* 73.3 (1980), pp. 1064–1072.

[110] R. Fukuda, M. Ehara, and H. Nakatsuji. "Excited states and electronic spectra of extended tetraazaporphyrins". In: *The Journal of Chemical Physics* 133.14 (2010), p. 144316.

[111] C. Adamo and V. Barone. "Toward reliable density functional methods without adjustable parameters: The PBE0 model". In: *The Journal of Chemical Physics* 110.13 (1999), pp. 6158–6170.

[112] F. Weigend and R. Ahlrichs. "Balanced basis sets of split valence, triple zeta valence and quadruple zeta valence quality for H to Rn: Design and assessment of accuracy". In: *Physical Chemistry Chemical Physics* 7.18 (2005), pp. 3297–3305.

[113] S. G. Balasubramani, G. P. Chen, S. Coriani, M. Diedenhofen, M. S. Frank, Y. J. Franzke, F. Furche, R. Grotjahn, M. E. Harding, C. Hättig, et al. "TURBOMOLE: Modular program suite for ab initio quantum-chemical and condensed-matter simulations". In: *The Journal of Chemical Physics* 152.18 (2020), p. 184107.

[114] Y. Baeten, E. Fron, C. Ruzié, Y. H. Geerts, and M. Van Der Auweraer. "Investigation of the Qx–Qy Equilibrium in a Metal-Free Phthalocyanine". In: *ChemPhysChem* 16.18 (2015), pp. 3992–3996.

[115] C. Chen, P. Chu, C. Bobisch, D. Mills, and W. Ho. "Viewing the interior of a single molecule: vibronically resolved photon imaging at submolecular resolution". In: *Physical Review Letters* 105.21 (2010), p. 217402.

[116] T. Neuman, R. Esteban, D. Casanova, F. J. García-Vidal, and J. Aizpurua. "Coupling of molecular emitters and plasmonic cavities beyond the point-dipole approximation". In: *Nano Letters* 18.4 (2018), pp. 2358–2364.

[117] P. Liljeroth, J. Repp, and G. Meyer. "Current-induced hydrogen tautomerization and conductance switching of naphthalocyanine molecules". In: *Science* 317.5842 (2007), pp. 1203–1206.

[118] J. Kügel, A. Sixta, M. Böhme, A. Krönlein, and M. Bode. "Breaking Degeneracy of Tautomerization-Metastability from Days to Seconds". In: *ACS Nano* 10.12 (2016), pp. 11058–11065.

[119] J. Kügel, M. Leisegang, M. Böhme, A. Krönlein, A. Sixta, and M. Bode. "Remote single-molecule switching: Identification and nanoengineering of hot electron-induced tautomerization". In: *Nano Letters* 17.8 (2017), pp. 5106–5112.

[120] T. Kumagai, F. Hanke, S. Gawinkowski, J. Sharp, K. Kotsis, J. Waluk, M. Persson, and L. Grill. "Thermally and vibrationally induced tautomerization of single porphycene molecules on a Cu (110) surface". In: *Physical Review Letters* 111.24 (2013), p. 246101.

[121] H. Böckmann, S. Gawinkowski, J. Waluk, M. B. Raschke, M. Wolf, and T. Kumagai. "Near-field enhanced photochemistry of single molecules in a scanning tunneling microscope junction". In: *Nano Letters* 18.1 (2018), pp. 152–157.

[122] J. N. Ladenthin, T. Frederiksen, M. Persson, J. C. Sharp, S. Gawinkowski, J. Waluk, and T. Kumagai. "Force-induced tautomerization in a single molecule". In: *Nature Chemistry* 8.10 (2016), pp. 935–940.

[123] J. G. Simmons. "Generalized formula for the electric tunnel effect between similar electrodes separated by a thin insulating film". In: *Journal of Applied Physics* 34.6 (1963), pp. 1793–1803.

[124] Y.-m. Kuang, Y.-j. Yu, Y. Luo, J.-z. Zhu, Y. Liao, Y. Zhang, and Z.-c. Dong. "Tunneling electron induced fluorescence from single porphyrin molecules decoupled by striped-phase octanethiol self-assembled monolayer". In: *Chinese Journal of Chemical Physics* 29.2 (2016), p. 157.

[125] M. Lindner, M. Valášek, M. Mayor, T. Frauhammer, W. Wulfhekel, and L. Gerhard. "Molecular graph paper". In: *Angewandte Chemie International Edition* 56.28 (2017), pp. 8290–8294.

[126] M. Lindner, M. Valášek, M. Mayor, T. Frauhammer, W. Wulfhekel, and L. Gerhard. "Molekulares Kästchenpapier". In: *Angewandte Chemie* 129.28 (2017), pp. 8405–8410.

[127] L. Gerhard, K. Edelmann, J. Homberg, M. Valášek, S. G. Bahoosh, M. Lukas, F. Pauly, M. Mayor, and W. Wulfhekel. "An electrically actuated molecular toggle switch". In: *Nature Communications* 8.1 (2017), pp. 1–10.

[128] L. Gerhard and M. Valášek. "Switching Behavior of Tripodal Molecules on Au (111) Studied With STM". In: *Encyclopedia of Interfacial Chemistry* (2018), pp. 271–280.

[129] T. Frauhammer, L. Gerhard, K. Edelmann, M. Lindner, M. Valášek, M. Mayor, and W. Wulfhekel. "Addressing a lattice of rotatable molecular dipoles with the electric field of an STM tip". In: *Physical Chemistry Chemical Physics* 23.8 (2021), pp. 4874–4881.

[130] X. Tao, Z. Dong, J. Yang, Y. Luo, J. Hou, and J. Aizpurua. "Influence of a dielectric layer on photon emission induced by a scanning tunneling microscope". In: *The Journal of Chemical Physics* 130.8 (2009), p. 084706.

[131] M. Lindner, M. Valášek, J. Homberg, K. Edelmann, L. Gerhard, W. Wulfhekel, O. Fuhr, T. Wächter, M. Zharnikov, V. Kolivoška, et al. "Importance of the Anchor Group Position (Para versus Meta) in Tetraphenylmethane Tripods: Synthesis and Self-Assembly Features". In: *Chemistry–A European Journal* 22.37 (2016), pp. 13218–13235.

[132] Y. Li Huang, Y. Lu, T. C. Niu, H. Huang, S. Kera, N. Ueno, A. T. S. Wee, and W. Chen. "Reversible Single-Molecule Switching in an Ordered Monolayer Molecular Dipole Array". In: *Small* 8.9 (2012), pp. 1423–1428.

[133] J. L. Zhang, J. Q. Zhong, J. D. Lin, W. P. Hu, K. Wu, G. Q. Xu, A. T. Wee, and W. Chen. "Towards single molecule switches". In: *Chemical Society Reviews* 44.10 (2015), pp. 2998–3022.

[134] J. Coombs, J. Gimzewski, B. Reihl, J. Sass, and R. Schlittler. "Photon emission experiments with the scanning tunnelling microscope". In: *Journal of Microscopy* 152.2 (1988), pp. 325–336.

[135] T. Yanai, D. P. Tew, and N. C. Handy. "A new hybrid exchange–correlation functional using the Coulomb-attenuating method (CAM-B3LYP)". In: *Chemical Physics Letters* 393.1 (2004), pp. 51–57. ISSN: 0009-2614.

[136] C. Holzer. "An improved seminumerical Coulomb and exchange algorithm for properties and excited states in modern density functional theory". In: *The Journal of Chemical Physics* 153.18 (2020), p. 184115.

[137] G. Tian, J.-C. Liu, and Y. Luo. "Density-matrix approach for the electroluminescence of molecules in a scanning tunneling microscope". In: *Physical Review Letters* 106.17 (2011), p. 177401.

[138] R. L. Martin. "Natural transition orbitals". In: *The Journal of Chemical Physics* 118.11 (2003), pp. 4775–4777.

[139] M. Valášek, M. Lindner, and M. Mayor. "Rigid multipodal platforms for metal surfaces". In: *Beilstein Journal of Nanotechnology* 7.1 (2016), pp. 374–405.

[140] S. Wu, G. Nazin, and W. Ho. "Intramolecular photon emission from a single molecule in a scanning tunneling microscope". In: *Physical Review B* 77.20 (2008), p. 205430.

[141] M. Chong. "Electrically driven fluorescence of single molecule junctions". Theses. Université de Strasbourg, 2016.

[142] P. E. Barclay, C. Santori, K.-M. Fu, R. G. Beausoleil, and O. Painter. "Coherent interference effects in a nano-assembled diamond NV center cavity-QED system". In: *Optics Express* 17.10 (2009), pp. 8081–8197.

[143] B. Luk'yanchuk, N. I. Zheludev, S. A. Maier, N. J. Halas, P. Nordlander, H. Giessen, and C. T. Chong. "The Fano resonance in plasmonic nanostructures and metamaterials". In: *Nature Materials* 9.9 (2010), pp. 707–715.

[144] A. E. Miroshnichenko, S. Flach, and Y. S. Kivshar. "Fano resonances in nanoscale structures". In: *Review of Modern Physics* 82 (3 2010), pp. 2257–2298.

[145] K. Miwa, M. Sakaue, and H. Kasai. "Effects of interference between energy absorption processes of molecule and surface plasmons on light emission induced by scanning tunneling microscopy". In: *Journal of the Physical Society of Japan* 82.12 (2013), p. 124707.

[146] Y. Vardi, E. Cohen-Hoshen, G. Shalem, and I. Bar-Joseph. "Fano resonance in an electrically driven plasmonic device". In: *Nano Letters* 16.1 (2016), pp. 748–752.

[147] S. Triebwasser, E. S. Dayhoff, and W. E. Lamb Jr. "Fine structure of the hydrogen atom. V". In: *Physical Review* 89.1 (1953), p. 98.

[148] A. Fragner, M. Göppl, J. Fink, M. Baur, R. Bianchetti, P. Leek, A. Blais, and A. Wallraff. "Resolving vacuum fluctuations in an electrical circuit by measuring the Lamb shift". In: *Science* 322.5906 (2008), pp. 1357–1360.

[149] M. V. Rybin, S. F. Mingaleev, M. F. Limonov, and Y. S. Kivshar. "Purcell effect and Lamb shift as interference phenomena". In: *Scientific Reports* 6.1 (2016), pp. 1–9.

[150] T. Uemura, M. Furumoto, T. Nakano, M. Akai-Kasaya, A. Saito, M. Aono, and Y. Kuwahara. "Local-plasmon-enhanced up-conversion fluorescence from copper phthalocyanine". In: *Chemical Physics Letters* 448.4-6 (2007), pp. 232–236.

[151] T. Uemura, M. Akai-Kasaya, A. Saito, M. Aono, and Y. Kuwahara. "Spatially resolved detection of plasmon-enhanced fluorescence using scanning tunneling microscopy". In: *Surface and Interface Analysis: An International Journal devoted*

to the development and application of techniques for the analysis of surfaces, interfaces and thin films 40.6-7 (2008), pp. 1050–1053.

[152] H. Liu, R. Nishitani, T. Han, Y. Ie, Y. Aso, and H. Iwasaki. "STM fluorescence of porphyrin enhanced by a strong plasmonic field and its nanoscale confinement in an STM cavity". In: *Physical Review B* 79.12 (2009), p. 125415.

[153] A. Okada, K. Kanazawa, K. Hayashi, N. Okawa, T. Kurita, O. Takeuchi, and H. Shigekawa. "Anomalous Light Emission from Metal Phthalocyanine Films on Au (111) Activated by Tunneling-Current-Induced Surface Plasmon". In: *Applied Physics Express* 3.1 (2010), p. 015201.

[154] A. Fujiki, Y. Miyake, Y. Oshikane, M. Akai-Kasaya, A. Saito, and Y. Kuwahara. "STM-induced light emission from thin films of perylene derivatives on the HOPG and Au substrates". In: *Nanoscale Research Letters* 6.1 (2011), pp. 1–8.

[155] J. Chen and J. X. Zhao. "Upconversion nanomaterials: synthesis, mechanism, and applications in sensing". In: *Sensors* 12.3 (2012), pp. 2414–2435.

[156] M. Kasha. "Characterization of electronic transitions in complex molecules". In: *Discussions of the Faraday Society* 9 (1950), pp. 14–19.

[157] K. Rebane and P. Saari. "Hot luminescence and relaxation processes in resonant secondary emission of solid matter". In: *Journal of Luminescence* 16.3 (1978), pp. 223–243.

[158] M. C. Chong, L. Sosa-Vargas, H. Bulou, A. Boeglin, F. Scheurer, F. Mathevet, and G. Schull. "Ordinary and hot electroluminescence from single-molecule devices: Controlling the emission color by chemical engineering". In: *Nano Letters* 16.10 (2016), pp. 6480–6484.

[159] M. S. Anderson. "Locally enhanced Raman spectroscopy with an atomic force microscope". In: *Applied Physics Letters* 76.21 (2000), pp. 3130–3132.

[160] N. Hayazawa, Y. Inouye, Z. Sekkat, and S. Kawata. "Metallized tip amplification of near-field Raman scattering". In: *Optics Communications* 183.1-4 (2000), pp. 333–336.

[161] B. Pettinger, G. Picardi, R. Schuster, and G. Ertl. "Surface enhanced Raman spectroscopy: towards single molecule spectroscopy". In: *Electrochemistry* 68.12 (2000), pp. 942–949.

[162] R. M. Stöckle, Y. D. Suh, V. Deckert, and R. Zenobi. "Nanoscale chemical analysis by tip-enhanced Raman spectroscopy". In: *Chemical Physics Letters* 318.1-3 (2000), pp. 131–136.

[163] C. Chen, N. Hayazawa, and S. Kawata. “A 1.7 nm resolution chemical analysis of carbon nanotubes by tip-enhanced Raman imaging in the ambient”. In: *Nature Communications* 5.1 (2014), pp. 1–5.

[164] J. S. Seldenthuis, H. S. van der Zant, M. A. Ratner, and J. M. Thijssen. “Electroluminescence spectra in weakly coupled single-molecule junctions”. In: *Physical Review B* 81.20 (2010), p. 205430.

[165] G. Tian and Y. Luo. “Electroluminescence of molecules in a scanning tunneling microscope: Role of tunneling electrons and surface plasmons”. In: *Physical Review B* 84.20 (2011), p. 205419.

[166] Y. Zhang, Y. Zelinskyy, and V. May. “Plasmon-enhanced electroluminescence of a single molecule: A theoretical study”. In: *Physical Review B* 88.15 (2013), p. 155426.

[167] K. Miwa, M. Sakaue, and H. Kasai. “Vibration-assisted upconversion of molecular luminescence induced by scanning tunneling microscopy”. In: *Nanoscale Research Letters* 8.1 (2013), pp. 1–6.

[168] G. Chen, X.-G. Li, Z.-Y. Zhang, and Z.-C. Dong. “Molecular hot electroluminescence due to strongly enhanced spontaneous emission rates in a plasmonic nanocavity”. In: *Nanoscale* 7.6 (2015), pp. 2442–2449.

[169] F. Silly and F. Charra. “Time-autocorrelation in scanning-tunneling-microscope-induced photon emission from metallic surface”. In: *Applied Physics Letters* 77.22 (2000), pp. 3648–3650.

[170] F. Silly and F. Charra. “Time-correlations as a contrast mechanism in scanning-tunneling-microscopy-induced photon emission”. In: *Ultramicroscopy* 99.2-3 (2004), pp. 159–164.

[171] K. Perronet, G. Schull, P. Raimond, and F. Charra. “Single-molecule fluctuations in a tunnel junction: A study by scanning-tunnelling-microscopy–induced luminescence”. In: *EPL (Europhysics Letters)* 74.2 (2006), p. 313.

[172] P. Merino, C. Große, A. Rosławska, K. Kuhnke, and K. Kern. “Exciton dynamics of C 60-based single-photon emitters explored by Hanbury Brown–Twiss scanning tunnelling microscopy”. In: *Nature Communications* 6.1 (2015), pp. 1–6.

[173] A. Rosławska, C. C. Leon, A. Grewal, P. Merino, K. Kuhnke, and K. Kern. “Atomic-scale dynamics probed by photon correlations”. In: *ACS Nano* 14.6 (2020), pp. 6366–6375.

Acknowledgments

I am glad to say that the last four years were the best time of my academic career. When I started my PhD, I was nervous about the fact that I did not know anyone in Karlsruhe, however, this feeling was very short lived and soon this place became home. Over the years, I got to know some very exciting people from whom I learned a lot. Here I like to present my humble gratitude and thank to everyone who supported me, helped me and made my life easier and exciting:

First and foremost, I would like to thank **Prof. Dr. Wulf Wulfhekel**. He trusted me and gave me the opportunity to work with him. It goes without saying that my PhD would not have turned out the way it did without his constant support and motivation. I feel very lucky that I joined PhD under his supervision. He is not only a great supervisor but also a great human being. I was amazed that how much he cared for all of us and not just professionally.

I would like to thank **Prof. Dr. David Hunger** for taking the time and acting as a co-referee.

I am really thankful to the financial support by **Deutscher Akademischer Austauschdienst** (DAAD).

A special thanks goes to **Dr. Lukas Gerhard** who was willing to help me unconditionally. I have to say that he was the best mentor one can hope for, also he is one of the most humble people I know. I learned a lot from him and not just professionally but also on personal levels. I really enjoyed our long scientific and nonscientific discussions. I have to say that his support and friendship means a lot to me.

I would like to thank our long term collaborators **Prof. Dr. Marcel Mayor, Dr. Michal Valášek** and **Dr. Nico Balzer** for providing the series of exciting new molecules, which ultimately resulted in the successful STML experiments discussed in this thesis.

I would like to thank **Prof. Dr. Carsten Rockstuhl, Dr. Christof Holzer, Dr. Taavi Repän,** and **Dr. Marjan Krstic** for the theoretical calculations, which really helped us to understand the light emission processes from different molecular moieties.

I would like to thank **Prof. Dr. Martin Wegener, Dr. Martin Schumann, Dr. Christian Kern** and **Dr. Liang Yang** for providing mirror tips and constantly working to improve their performance.

I would like to thank **Simone Dehm** and **Min-ken Li** for doing the SEM imaging of the mirror tips on short notices and **Dr. Thomas Reisinger** for introducing me to the laser cutting setup.

I am grateful to **Dr. Kevin Edelmann, Lars Wilmes, Qing Sun** and **Gabriel Derenbach**, who helped me at a different points of time during my PhD. A special thanks goes to **Dr. Kevin Edelmann** for building such a nice setup of γ-STM and introducing me to the lab.

I would like to thank all past and present group members of **AG Wulf Wulfhekel** for making my life pleasant. This includes **Julian, Timo, Loïc, Thomas, Marie, Fang, Jin-jie, Timofey, Jasmin, Namrata, Hongyan, Hung-Hsiang, Qili, Philip, Mirjam, Anika, Avisek, Lennart, Melvin and Paul**.

It goes without saying that this journey would not have turned out to be as much fun as it did without my family and friends **Vikas Rai, Nandini Rai, Sandeep Kumar, Navid Hussain, Devang Parmar, Neha Chaudhary, Subodh Sahoo and Sumit Kediyal**. I enjoyed the time spent with them over the years.

Finally, I would like to thank the three very important people in my life, my father, **Arvind Rai**, my sister, **Vishakha Rai** and my partner, **Neha Jha**. Without their constant support and motivation, this journey would not have been possible. I am very grateful to have **Neha** in my life, she was always with me, and tolerated me, no matter the situation.

Experimental Condensed Matter Physics
(ISSN 2191-9925)

Band 1 Alexey Feofanov
Experiments on flux qubits with pi-shifters.
ISBN 978-3-86644-644-1

Band 2 Stefan Schmaus
Spintronics with individual metal-organic molecules.
ISBN 978-3-86644-649-6

Band 3 Marc Müller
Elektrischer Leitwert von magnetostriktiven Dy-Nanokontakten.
ISBN 978-3-86644-726-4

Band 4 Torben Peichl
Einfluss mechanischer Deformation auf atomare Tunnelsysteme – untersucht mit Josephson Phasen-Qubits.
ISBN 978-3-86644-837-7

Band 5 Dominik Stöffler
Herstellung dünner metallischer Brücken durch Elektromigration und Charakterisierung mit Rastersondentechniken.
ISBN 978-3-86644-843-8

Band 6 Tihomir Tomanic
Untersuchung des elektronischen Oberflächenzustands von Ag-Inseln auf supraleitendem Niob (110).
ISBN 978-3-86644-898-8

Band 7 Lukas Gerhard
Magnetoelectric coupling at metal surfaces.
ISBN 978-3-7315-0063-6

Band 8 Nicht erschienen.

Band 9 Jochen Zimmer
Cooper pair transport in arrays of Josephson junctions.
ISBN 978-3-7315-0130-5

Band 10 Oliver Berg
Elektrischer Transport durch Nanokontakte von Selten-Erd-Metallen.
ISBN 978-3-7315-0209-8

Band 11 Grigorij Jur'evic Grabovskij
Investigation of coherent microscopic defects inside the tunneling barrier of a Josephson junction.
ISBN 978-3-7315-0210-4

Die Bände sind unter www.ksp.kit.edu als PDF frei verfügbar oder als Druckausgabe bestellbar.

Band 12	Cornelius Thiele **STM Characterization of Phenylene-Ethynylene Oligomers on Au(111) and their Integration into Carbon Nanotube Nanogaps.** ISBN 978-3-7315-0235-7
Band 13	Michael Peter Schackert **Scanning Tunneling Spectroscopy on Electron-Boson Interactions in Superconductors.** ISBN 978-3-7315-0238-8
Band 14	Susanne Butz **One-Dimensional Tunable Josephson Metamaterials.** ISBN 978-3-7315-0271-5
Band 15	Philipp Jung **Nonlinear Effects in Superconducting Quantum Interference Meta-Atoms.** ISBN 978-3-7315-0294-4
Band 16	Sebastian Probst **Hybrid quantum system based on rare earth doped crystals.** ISBN 978-3-7315-0345-3
Band 17	Wolfram Kittler **Magnetische Anisotropie und Quantenphasenübergang in $CeTi_{1-x}V_xGe_3$.** ISBN 978-3-7315-0363-7
Band 18	Moritz Peter **Towards magnetic resonance in scanning tunneling microscopy using heterodyne detection.** ISBN 978-3-7315-0410-8
Band 19	Junji Tobias Märkl **Investigation of Magnetic Adatoms with Scanning Tunneling Techniques.** ISBN 978-3-7315-0435-1
Band 20	Jochen Braumüller **Quantum simulation experiments with superconducting circuits.** ISBN 978-3-7315-0780-2
Band 21	Jinjie Chen **Local Investigation of Single Magnetic Molecules with Scanning Tunneling Microscopy.** ISBN 978-3-7315-0819-9
Band 22	Arnold Seiler **Einfluss der Leitungselektronen auf die Dynamik atomarer Tunnelsysteme in ungeordneten Festkörpern: Relaxationsprozesse in metallischen Gläsern und ungeordneten dünnen Aluminiumoxid-Schichten.** ISBN 978-3-7315-0870-0

Band 23 Jasmin Maria Jandke
Elastic and Inelastic Scanning Tunneling Spectroscopy on Iron-Based Superconductors.
ISBN 978-3-7315-0747-5

Band 24 Kevin Edelmann
Electroluminescence from Plasmonic Excitations in a Scanning Tunnelling Microscope.
ISBN 978-3-7315-0923-3

Band 25 Lei Zhang
Sub-Kelvin scanning tunneling microscopy on magnetic molecules.
ISBN 978-3-86644-950-3

Band 26 Lukas Grünhaupt
Granular aluminium superinductors.
ISBN 978-3-7315-0956-1

Band 27 Alexander Bilmes
Resolving locations of defects in superconducting transmon qubits.
ISBN 978-3-7315-0967-7

Band 28 Marco Pfirrmann
Adding nonlinearity to an electromagnetic-magnonic quantum hybrid device.
ISBN 978-3-7315-1003-1

Band 29 Andre Schneider
Quantum Sensing Experiments with Superconducting Qubits.
ISBN 978-3-7315-1032-1

Band 30 Vibhuti Narayan Rai
Light Emission from Single Self-decoupled Molecules in a Scanning Tunnelling Microscope.
ISBN 978-3-7315-1201-1